# 対訳
# ISO
# 14001:2015
（JIS Q 14001:2015）

**ポケット版**

# 環境マネジメントの
# 国際規格

日本規格協会　編

2019 年 7 月 1 日の JIS 法改正により名称が変わりました．本書に収録している JIS についても，まえがきを除き，規格中の「日本工業規格」を「日本産業規格」に読み替えてください．

\*著作権について

　本書は，ISO 中央事務局と当会との翻訳出版契約に基づいて刊行したものです．

　本書に収録した ISO 及び JIS は，著作権により保護されています．本書の一部又は全部について，当会及び ISO の許可なく複写・複製することを禁じます．ISO の著作権は，下に示すとおりです．

　本書の著作権に関するお問い合わせは，日本規格協会グループ（e-mail：copyright@jsa.or.jp）にて承ります．

---

© ISO 2015

All rights reserved. Unless otherwise specified, no part of this publication may be reproduced or utilized otherwise in any form or by any means, electronic or mechanical, including photocopying, or posting on the internet or an intranet, without prior written permission. Permission can be requested from either ISO at the address below or ISO's member body in the country of the requester.

　ISO copyright office

　Ch. de Blandonnet 8・CP 401

　CH-1214 Vernier, Geneva, Switzerland

　Tel. +41 22 749 01 11

　Fax +41 22 749 09 47

　E-mail copyright@iso.org

　Web www.iso.org

# まえがき

環境マネジメントシステムの国際規格として1996 年に制定された ISO 14001 は，2004 年の改訂を経て，2015 年 9 月 15 日に第 3 版として改訂された．

このたびの改訂では，ISO 9001 をはじめ，他のマネジメントシステム規格との整合化のため，ISO/IEC 専門業務用指針　第 1 部　統合版 ISO 補足指針　附属書 SL に基づく共通の規格構造，要求事項，用語・定義が採用された．

これにより規格構造上の変更が生じたとともに，戦略的な環境マネジメント，リスク及び機会，トップマネジメントのリーダーシップに対する責任の強化をはじめとする，新たな要求事項が追加された．

改訂ではまた，組織を取り巻く状況や環境課題・技術の，旧版発行以降の変化に対応するため，環境パフォーマンスの改善の重視，ライフサイクル思考の導入などを組み込んだ要求事項の見直しも行われた．

このように，旧版から 11 年ぶりとなる今改訂は，現在及び今後の環境課題に対応できる環境マネ

ジメントシステム規格とするための抜本的な大改訂
となった.

　国内では，この国際規格の改訂を受けて，2014
年に環境管理システム小委員会において，ISO
14001 を基に JIS Q 14001 を改正することを決定
した．そして，当該委員会とその下に設置された
ISO 14001 改正検討 WG において JIS 原案を作成
し，2015 年 11 月 20 日に JIS Q 14001:2015 の公
示・発行に至った.

　本書は，その ISO と JIS を対訳版として発刊す
るものである.
　英文は ISO 14001:2015 を ISO（国際標準化機
構）の許可を得て，収録している．また，和文は，
当該 ISO について技術的内容及び対応国際規格の
構成を変更することなく作成され，経済産業省の
日本工業標準調査会で審議を経て改正された JIS Q
14001:2015 を収録している．なお，JIS 規格文中
において点線下線を施した箇所は，原国際規格には
ない事項であること，また，JIS の解説は省略して
いることに留意いただきたい.

　このたびの ISO 14001:2015 の規格開発，及び

JIS Q 14001 の 2015 年改正にあたり，環境管理システム小委員会及び ISO 14001 改正検討 WG の委員各位には，国内外の会議での検討など，多大なご尽力をいただいた．ここに厚く感謝の意を表する．

本書は，環境マネジメントシステムの更新審査を受ける企業，あるいはこれから新たに審査登録を受ける企業のための原典として，前版同様，お役に立てれば幸いである．また，これらの規格をより深く理解したい方に対しては，書籍『ISO 14001:2015 要求事項の解説』，『ISO 14001:2015 新旧規格の対照と解説』（いずれも，日本規格協会，2015）をはじめとする関係書籍を併読されることをお勧めする．

　2016 年 2 月

日本規格協会

# 環境管理システム小委員会

(2015 年 11 月現在)

[委員長] 吉田　敬史[*]　合同会社グリーンフューチャーズ
[委　員] 池田　真夫　一般社団法人日本貿易会
　　　　　　　　　　（2015 年 3 月 31 日まで）

　　　　　中西　正治　一般社団法人日本貿易会
　　　　　　　　　　（2015 年 4 月 1 日から）

　　　　　伊吹　隆直　一般社団法人日本鉄鋼連盟
　　　　　大熊　一寛　環境省（2015 年 7 月 31 日まで）
　　　　　奥山　祐矢　環境省（2015 年 8 月 1 日から）
　　　　　奥野麻衣子[*]　三菱 UFJ リサーチ＆コンサルティング
　　　　　　　　　　株式会社

　　　　　小原愼一郎[*]　小原 MSC 事務所
　　　　　吉良　雅治　一般社団法人日本産業機械工業会
　　　　　小山　敏明　一般社団法人日本建設業連合会
　　　　　　　　　　（清水建設株式会社）

　　　　　笹森　幹雄[*]　日本マネジメントシステム認証機関協議会
　　　　　　　　　　（一般社団法人日本能率協会）

　　　　　澤本　雅人　審査員研修機関連絡協議会
　　　　　　　　　　（株式会社グローバルテクノ）

　　　　　下川　祐太　一般社団法人産業環境管理協会
　　　　　　　　　　環境マネジメントシステム審査員評価登録
　　　　　　　　　　センター（2015 年 3 月 31 日まで）

　　　　　鈴木　義仁　一般社団法人産業環境管理協会
　　　　　　　　　　環境マネジメントシステム審査員評価登録
　　　　　　　　　　センター（2015 年 4 月 1 日から）

　　　　　寺田　　博[*]　IMS コンサルティング株式会社
　　　　　野中　玲子　一般社団法人日本化学工業協会

|          | 平林　良人 | ISO/TC176 国内委員会 |
|----------|-----------|---------------------|
|          |           | （株式会社テクノファ） |
|          | 福田　泰和 | 経済産業省 |
|          | 細野　貴靖 | 東京電力株式会社（2015 年 6 月 30 日まで） |
|          | 岡峰　克幸 | 電気事業連合会（2015 年 7 月 1 日から） |
|          | 牧野　睦子* | 公益財団法人日本適合性認定協会 |
|          | 山田　秀 | 筑波大学 |
| ［関係者］ | 藤代　尚武 | 経済産業省 |
|          | 宮尾　健* | 経済産業省 |
|          | 佐々木千晶* | 経済産業省（2014 年 11 月 30 日まで） |
|          | 岡崎　将* | 経済産業省 |
|          | 谷口　淳子* | 経済産業省（2015 年 7 月 6 日まで） |
|          | 小林　麻子* | 経済産業省（2015 年 7 月 7 日から） |
|          | 齋藤　優子* | 経済産業省 |
|          | 齋藤　英亜* | 環境省 |
| ［事務局］ | 佐藤　恭子* | 一般財団法人日本規格協会 |
|          | 山下　岩男* | 一般財団法人日本規格協会 |
|          |           | （2015 年 5 月 19 日まで） |
|          | 遠藤　智之* | 一般財団法人日本規格協会 |
|          |           | （2015 年 5 月 20 日から） |
|          | 高井　玉歩* | 一般財団法人日本規格協会 |

注記　*印は，分科会委員を示す．

（敬称略）

# ISO 14001 改正検討 WG

(2015 年 11 月現在)

| | | |
|---|---|---|
| ［主　査］ | 吉田　敬史 | 合同会社グリーンフューチャーズ |
| ［委　員］ | 奥野麻衣子 | 三菱 UFJ リサーチ & コンサルティング 株式会社 |
| | 小原愼一郎 | 小原 MSC 事務所 |
| | 川口　努 | 富士通株式会社 |
| | 笹森　幹雄 | 日本マネジメントシステム認証機関協議会 （一般社団法人日本能率協会） |
| | 須田　晋介 | ISO/TC176 国内委員会 （株式会社テクノファ） |
| | 高戸　満 | 一般社団法人産業環境管理協会 環境マネジメントシステム審査員評価登録 センター |
| | 寺田　博 | IMS コンサルティング株式会社 |
| | 中ノ　理子 | イオン株式会社 |
| | 牧野　睦子 | 公益財団法人日本適合性認定協会 |
| ［関係者］ | 宮尾　健 | 経済産業省 |
| | 佐々木千晶 | 経済産業省（2014 年 11 月 30 日まで） |
| | 岡崎　将 | 経済産業省 |
| | 谷口　淳子 | 経済産業省（2015 年 7 月 6 日まで） |
| | 小林　麻子 | 経済産業省（2015 年 7 月 7 日から） |
| | 齋藤　優子 | 経済産業省 |
| | 齋藤　英亜 | 環境省 |

| ［事務局］ | 佐藤　恭子 | 一般財団法人日本規格協会 |
| | 山下　岩男 | 一般財団法人日本規格協会 |
| | | （2015 年 5 月 19 日まで） |
| | 遠藤　智之 | 一般財団法人日本規格協会 |
| | | （2015 年 5 月 20 日から） |
| | 高井　玉歩 | 一般財団法人日本規格協会 |

（敬称略）

## Contents

## ISO 14001:2015
## Environmental management systems
## — Requirements with guidance for use

Foreword ･･････････････････････････････････････ 20

Introduction ･･････････････････････････････････ 26

0.1　Background ･････････････････････････ 26

0.2　Aim of an environmental management
　　　system ････････････････････････････ 28

0.3　Success factors ･･･････････････････････ 30

0.4　Plan-Do-Check-Act model ･･･････････ 34

0.5　Contents of this International Standard
　　　････････････････････････････････････ 38

1　　Scope ･･････････････････････････････ 44

2　　Normative references ･･･････････････ 48

3　　Terms and definitions ･･････････････ 48

3.1　Terms related to organization and
　　　leadership ･････････････････････････ 48

3.2　Terms related to planning ･･････････ 54

3.3　Terms related to support and operation ･･･ 68

3.4　Terms related to performance evaluation
　　　and improvement ･･･････････････････ 72

# 目　次

## JIS Q 14001:2015
## 環境マネジメントシステム
## ―要求事項及び利用の手引

| | | |
|---|---|---|
| まえがき | …………………………………… | 21 |
| 序文 | ……………………………………… | 27 |
| 0.1 | 背景 ………………………………… | 27 |
| 0.2 | 環境マネジメントシステムの狙い ………… | 29 |
| 0.3 | 成功のための要因 ………………………… | 31 |
| 0.4 | Plan-Do-Check-Act モデル ……………… | 35 |
| 0.5 | この規格の内容 …………………………… | 39 |
| 1 | 適用範囲 …………………………………… | 45 |
| 2 | 引用規格 …………………………………… | 49 |
| 3 | 用語及び定義 ……………………………… | 49 |
| 3.1 | 組織及びリーダーシップに関する用語 …… | 49 |
| 3.2 | 計画に関する用語 ………………………… | 55 |
| 3.3 | 支援及び運用に関する用語 ……………… | 69 |
| 3.4 | パフォーマンス評価及び改善に関する用語 | |
| | …………………………………… | 73 |

| 4 | Context of the organization | 82 |
|---|---|---|
| 4.1 | Understanding the organization and its context | 82 |
| 4.2 | Understanding the needs and expectations of interested parties | 82 |
| 4.3 | Determining the scope of the environmental management system | 84 |
| 4.4 | Environmental management system | 86 |
| 5 | Leadership | 86 |
| 5.1 | Leadership and commitment | 86 |
| 5.2 | Environmental policy | 90 |
| 5.3 | Organizational roles, responsibilities and authorities | 92 |
| 6 | Planning | 94 |
| 6.1 | Actions to address risks and opportunities | 94 |
| 6.1.1 | General | 94 |
| 6.1.2 | Environmental aspects | 96 |
| 6.1.3 | Compliance obligations | 100 |
| 6.1.4 | Planning action | 102 |
| 6.2 | Environmental objectives and planning to achieve them | 104 |
| 6.2.1 | Environmental objectives | 104 |

| | | |
|---|---|---|
| 4 | 組織の状況 | 83 |
| 4.1 | 組織及びその状況の理解 | 83 |
| 4.2 | 利害関係者のニーズ及び期待の理解 | 83 |
| 4.3 | 環境マネジメントシステムの適用範囲の決定 | 85 |
| 4.4 | 環境マネジメントシステム | 87 |
| 5 | リーダーシップ | 87 |
| 5.1 | リーダーシップ及びコミットメント | 87 |
| 5.2 | 環境方針 | 91 |
| 5.3 | 組織の役割，責任及び権限 | 93 |
| 6 | 計画 | 95 |
| 6.1 | リスク及び機会への取組み | 95 |
| 6.1.1 | 一般 | 95 |
| 6.1.2 | 環境側面 | 97 |
| 6.1.3 | 順守義務 | 101 |
| 6.1.4 | 取組みの計画策定 | 103 |
| 6.2 | 環境目標及びそれを達成するための計画策定 | 105 |
| 6.2.1 | 環境目標 | 105 |

| | | |
|---|---|---|
| 6.2.2 | Planning actions to achieve environmental objectives | 104 |
| 7 | Support | 106 |
| 7.1 | Resources | 106 |
| 7.2 | Competence | 106 |
| 7.3 | Awareness | 108 |
| 7.4 | Communication | 110 |
| 7.4.1 | General | 110 |
| 7.4.2 | Internal communication | 112 |
| 7.4.3 | External communication | 114 |
| 7.5 | Documented information | 114 |
| 7.5.1 | General | 114 |
| 7.5.2 | Creating and updating | 116 |
| 7.5.3 | Control of documented information | 116 |
| 8 | Operation | 120 |
| 8.1 | Operational planning and control | 120 |
| 8.2 | Emergency preparedness and response | 124 |
| 9 | Performance evaluation | 126 |
| 9.1 | Monitoring, measurement, analysis and evaluation | 126 |
| 9.1.1 | General | 126 |
| 9.1.2 | Evaluation of compliance | 130 |
| 9.2 | Internal audit | 132 |
| 9.2.1 | General | 132 |

| | | |
|---|---|---|
| 6.2.2 | 環境目標を達成するための取組みの計画策定 …………………………………… | 105 |
| 7 | 支援 ………………………………………… | 107 |
| 7.1 | 資源 ………………………………………… | 107 |
| 7.2 | 力量 ………………………………………… | 107 |
| 7.3 | 認識 ………………………………………… | 109 |
| 7.4 | コミュニケーション ……………………… | 111 |
| 7.4.1 | 一般 ………………………………………… | 111 |
| 7.4.2 | 内部コミュニケーション ………………… | 113 |
| 7.4.3 | 外部コミュニケーション ………………… | 115 |
| 7.5 | 文書化した情報 …………………………… | 115 |
| 7.5.1 | 一般 ………………………………………… | 115 |
| 7.5.2 | 作成及び更新 ……………………………… | 117 |
| 7.5.3 | 文書化した情報の管理 …………………… | 117 |
| 8 | 運用 ………………………………………… | 121 |
| 8.1 | 運用の計画及び管理 ……………………… | 121 |
| 8.2 | 緊急事態への準備及び対応 ……………… | 125 |
| 9 | パフォーマンス評価 ……………………… | 127 |
| 9.1 | 監視，測定，分析及び評価 ……………… | 127 |
| 9.1.1 | 一般 ………………………………………… | 127 |
| 9.1.2 | 順守評価 …………………………………… | 131 |
| 9.2 | 内部監査 …………………………………… | 133 |
| 9.2.1 | 一般 ………………………………………… | 133 |

9.2.2 Internal audit programme ·················· 132

9.3 Management review ························· 134

10 Improvement ··························· 138

10.1 General ······························· 138

10.2 Nonconformity and corrective action ··· 140

10.3 Continual improvement ················· 142

Annex A (informative)　Guidance on the use of

this International Standard ················ 144

Annex B (informative)　Correspondence between

ISO 14001:2015 and ISO 14001:2004 ···· 240

Bibliography ································· 248

Alphabetical index of terms（省略）

9.2.2 内部監査プログラム ……………………………… 133

9.3 マネジメントレビュー ………………………… 135

10 改善 ………………………………………… 139

10.1 一般 ……………………………………… 139

10.2 不適合及び是正処置 …………………………… 141

10.3 継続的改善 ……………………………………… 143

附属書A(参考) この規格の利用の手引 ……… 145

附属書B(参考) JIS Q 14001:2015 と JIS Q
14001:2004 との対応 ……………………… 241

参考文献 ……………………………………………… 249

索引

五十音順 ………………………………………… 255

アルファベット順 ……………………………… 260

**ISO 14001**
Third edition 2015–09–15

**JIS Q 14001**
2015–11–20

Environmental management systems
— Requirements with guidance for use

環境マネジメントシステム
—要求事項及び利用の手引

## Foreword

ISO (the International Organization for Standardization) is a worldwide federation of national standards bodies (ISO member bodies). The work of preparing International Standards is normally carried out through ISO technical committees. Each member body interested in a subject for which a technical committee has been established has the right to be represented on that committee. International organizations, governmental and non-governmental, in liaison with ISO, also take part in the work. ISO collaborates closely with the International Electrotechnical Commission (IEC) on all matters of electrotechnical standardization.

The procedures used to develop this document and those intended for its further maintenance are described in the ISO/IEC Directives, Part 1. In particular the different approval criteria needed for the different types of ISO documents should be noted. This document was drafted in accordance with the editorial rules of the ISO/IEC Directives,

# まえがき

(ISO の Foreword と JIS のまえがきは，それぞれの原文において内容が異なっているため，対訳となっていないことにご注意ください.)

この規格は，工業標準化法第 14 条によって準用する第 12 条第 1 項の規定に基づき，一般財団法人日本規格協会（JSA）から，工業標準原案を具して日本工業規格を改正すべきとの申出があり，日本工業標準調査会の審議を経て，経済産業大臣が改正した日本工業規格である．これによって，**JIS Q 14001**:2004 は改正され，この規格に置き換えられた．

この規格は，著作権法で保護対象となっている著作物である．

この規格の一部が，特許権，出願公開後の特許出願又は実用新案権に抵触する可能性があることに注意を喚起する．経済産業大臣及び日本工業標準調査会は，このような特許権，出願公開後の特許出願及び実用新案権に関わる確認について，責任はもたない．

Part 2 (see www.iso.org/directives).

Attention is drawn to the possibility that some of the elements of this document may be the subject of patent rights. ISO shall not be held responsible for identifying any or all such patent rights. Details of any patent rights identified during the development of the document will be in the Introduction and/or on the ISO list of patent declarations received (see www.iso.org/patents).

Any trade name used in this document is information given for the convenience of users and does not constitute an endorsement.

For an explanation on the meaning of ISO specific terms and expressions related to conformity assessment, as well as information about ISO's adherence to the World Trade Organization (WTO) principles in the Technical Barriers to Trade (TBT) see the following URL: www.iso.org/iso/foreword.html.

The committee responsible for this document is

23

Technical Committee ISO/TC 207, *Environmental management*, Subcommittee SC 1, *Environmental management systems*.

This third edition cancels and replaces the second edition (ISO 14001:2004), which has been technically revised. It also incorporates the Technical Corrigendum ISO 14001:2004/Cor.1:2009.

25

# Introduction

## 0.1 Background

Achieving a balance between the environment, society and the economy is considered essential to meet the needs of the present without compromising the ability of future generations to meet their needs. Sustainable development as a goal is achieved by balancing the three pillars of sustainability.

Societal expectations for sustainable development, transparency and accountability have evolved with increasingly stringent legislation, growing pressures on the environment from pollution, inefficient use of resources, improper waste management, climate change, degradation of ecosystems and loss of biodiversity.

序文

この規格は，2015 年に第 3 版として発行された
**ISO 14001** を基に，技術的内容及び構成を変更す
ることなく作成した日本工業規格である．

なお，この規格で点線の下線を施してある参考事
項は，対応国際規格にはない事項である．

## 0.1　背景

将来の世代の人々が自らのニーズを満たす能力を
損なうことなく，現在の世代のニーズを満たすため
に，環境，社会及び経済のバランスを実現すること
が不可欠であると考えられている．到達点としての
持続可能な開発は，持続可能性のこの "三本柱" の
バランスをとることによって達成される．

厳格化が進む法律，汚染による環境への負荷の増
大，資源の非効率的な使用，不適切な廃棄物管理，
気候変動，生態系の劣化及び生物多様性の喪失に伴
い，持続可能な開発，透明性及び説明責任に対する
社会の期待は高まっている．

This has led organizations to adopt a systematic approach to environmental management by implementing environmental management systems with the aim of contributing to the environmental pillar of sustainability.

## 0.2   Aim of an environmental management system

The purpose of this International Standard is to provide organizations with a framework to protect the environment and respond to changing environmental conditions in balance with socio-economic needs. It specifies requirements that enable an organization to achieve the intended outcomes it sets for its environmental management system.

A systematic approach to environmental management can provide top management with information to build success over the long term and create options for contributing to sustainable development by:

— protecting the environment by preventing or mitigating adverse environmental impacts;

— mitigating the potential adverse effect of envi-

こうしたことから，組織は，持続可能性の"環境の柱"に寄与することを目指して，環境マネジメントシステムを実施することによって環境マネジメントのための体系的なアプローチを採用するようになってきている．

## 0.2　環境マネジメントシステムの狙い

この規格の目的は，社会経済的ニーズとバランスをとりながら，環境を保護し，変化する環境状態に対応するための枠組みを組織に提供することである．この規格は，組織が，環境マネジメントシステムに関して設定する意図した成果を達成することを可能にする要求事項を規定している．

環境マネジメントのための体系的なアプローチは，次の事項によって，持続可能な開発に寄与することについて，長期的な成功を築き，選択肢を作り出すための情報を，トップマネジメントに提供することができる．

—　有害な環境影響を防止又は緩和することによって，環境を保護する．
—　組織に対する，環境状態から生じる潜在的で有

ronmental conditions on the organization;

— assisting the organization in the fulfilment of compliance obligations;

— enhancing environmental performance;

— controlling or influencing the way the organization's products and services are designed, manufactured, distributed, consumed and disposed by using a life cycle perspective that can prevent environmental impacts from being unintentionally shifted elsewhere within the life cycle;

— achieving financial and operational benefits that can result from implementing environmentally sound alternatives that strengthen the organization's market position;

— communicating environmental information to relevant interested parties.

This International Standard, like other International Standards, is not intended to increase or change an organization's legal requirements.

## 0.3  Success factors

The success of an environmental management sys-

序　文　　　　　31

害な影響を緩和する．

— 組織が順守義務を満たすことを支援する．

— 環境パフォーマンスを向上させる．
— 環境影響が意図せずにライフサイクル内の他の
部分に移行するのを防ぐことができるライフサ
イクルの視点を用いることによって，組織の製
品及びサービスの設計，製造，流通，消費及び
廃棄の方法を管理するか，又はこの方法に影響
を及ぼす．

— 市場における組織の位置付けを強化し，かつ，
環境にも健全な代替策を実施することで，財務
上及び運用上の便益を実現する．

— 環境情報を，関連する利害関係者に伝達する．

　この規格は，他の規格と同様に，組織の法的要求
事項を増大又は変更させることを意図していない．

## 0.3　成功のための要因
　環境マネジメントシステムの成功は，トップマネ

tem depends on commitment from all levels and functions of the organization, led by top management. Organizations can leverage opportunities to prevent or mitigate adverse environmental impacts and enhance beneficial environmental impacts, particularly those with strategic and competitive implications. Top management can effectively address its risks and opportunities by integrating environmental management into the organization's business processes, strategic direction and decision making, aligning them with other business priorities, and incorporating environmental governance into its overall management system. Demonstration of successful implementation of this International Standard can be used to assure interested parties that an effective environmental management system is in place.

Adoption of this International Standard, however, will not in itself guarantee optimal environmental outcomes. Application of this International Standard can differ from one organization to another due to the context of the organization. Two organizations can carry out similar activities but can

ジメントが主導する，組織の全ての階層及び機能からのコミットメントのいかんにかかっている．組織は，有害な環境影響を防止又は緩和し，有益な環境影響を増大させるような機会，中でも戦略及び競争力に関連のある機会を活用することができる．トップマネジメントは，他の事業上の優先事項と整合させながら，環境マネジメントを組織の事業プロセス，戦略的な方向性及び意思決定に統合し，環境上のガバナンスを組織の全体的なマネジメントシステムに組み込むことによって，リスク及び機会に効果的に取り組むことができる．この規格をうまく実施していることを示せば，有効な環境マネジメントシステムをもつことを利害関係者に確信させることができる．

　しかし，この規格の採用そのものが，最適な環境上の成果を保証するわけではない．この規格の適用は，組織の状況によって，各組織で異なり得る．二つの組織が，同様の活動を行っていながら，それぞれの順守義務，環境方針におけるコミットメント，環境技術及び環境パフォーマンスの到達点が異なる

have different compliance obligations, commitments in their environmental policy, environmental technologies and environmental performance goals, yet both can conform to the requirements of this International Standard.

The level of detail and complexity of the environmental management system will vary depending on the context of the organization, the scope of its environmental management system, its compliance obligations, and the nature of its activities, products and services, including its environmental aspects and associated environmental impacts.

## 0.4   Plan-Do-Check-Act model

The basis for the approach underlying an environmental management system is founded on the concept of Plan-Do-Check-Act (PDCA). The PDCA model provides an iterative process used by organizations to achieve continual improvement. It can be applied to an environmental management system and to each of its individual elements. It can be briefly described as follows.

—   Plan: establish environmental objectives and

場合であっても，共にこの規格の要求事項に適合することがあり得る．

環境マネジメントシステムの詳細さ及び複雑さのレベルは，組織の状況，環境マネジメントシステムの適用範囲，順守義務，並びに組織の活動，製品及びサービスの性質（これらの環境側面及びそれに伴う環境影響も含む．）によって異なる．

## 0.4 Plan-Do-Check-Act モデル

環境マネジメントシステムの根底にあるアプローチの基礎は，Plan-Do-Check-Act（PDCA）という概念に基づいている．PDCA モデルは，継続的改善を達成するために組織が用いる反復的なプロセスを示している．PDCA モデルは，環境マネジメントシステムにも，その個々の要素の各々にも適用できる．PDCA モデルは，次のように簡潔に説明できる．

― Plan： 組織の環境方針に沿った結果を出す

processes necessary to deliver results in accordance with the organization's environmental policy.

— Do: implement the processes as planned.

— Check: monitor and measure processes against the environmental policy, including its commitments, environmental objectives and operating criteria, and report the results.

— Act: take actions to continually improve.

Figure 1 shows how the framework introduced in this International Standard could be integrated into a PDCA model, which can help new and existing users to understand the importance of a systems approach.

ために必要な環境目標及びプロセスを確立する.

— Do： 計画どおりにプロセスを実施する.
— Check： コミットメントを含む環境方針，環境目標及び運用基準に照らして，プロセスを監視し，測定し，その結果を報告する.

— Act： 継続的に改善するための処置をとる.

図1は，この規格に導入された枠組みが，どのようにPDCAモデルに統合され得るかを示しており，新規及び既存の利用者がシステムアプローチの重要性を理解する助けとなり得る.

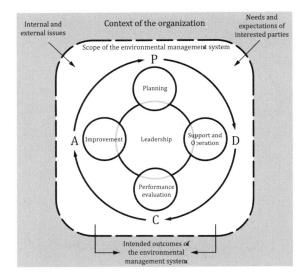

Figure 1 — Relationship between PDCA and the framework in this International Standard

## 0.5 Contents of this International Standard

This International Standard conforms to ISO's requirements for management system standards. These requirements include a high level structure, identical core text, and common terms with core definitions, designed to benefit users implementing multiple ISO management system standards.

This International Standard does not include re-

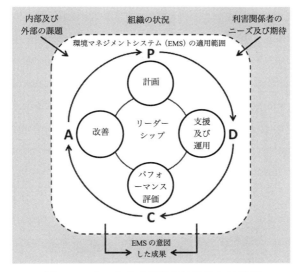

図1 - PDCA とこの規格の枠組みとの関係

## 0.5 この規格の内容

この規格は,国際標準化機構(ISO)及び JIS のマネジメントシステム規格に対する要求事項に適合している.これらの要求事項は,複数の ISO 及び JIS のマネジメントシステム規格を実施する利用者の便益のために作成された,上位構造,共通の中核となるテキスト,共通用語及び中核となる定義を含んでいる.

この規格には,品質マネジメント,労働安全衛生

quirements specific to other management systems, such as those for quality, occupational health and safety, energy or financial management. However, this International Standard enables an organization to use a common approach and risk-based thinking to integrate its environmental management system with the requirements of other management systems.

This International Standard contains the requirements used to assess conformity. An organization that wishes to demonstrate conformity with this International Standard can do so by:

— making a self-determination and self-declaration, or

— seeking confirmation of its conformance by parties having an interest in the organization, such as customers, or

— seeking confirmation of its self-declaration by a party external to the organization, or

— seeking certification/registration of its environmental management system by an external organization.

序　文　　　41

マネジメント，エネルギーマネジメント，財務マネ
ジメントなどの他のマネジメントシステムに固有な
要求事項は含まれていない．しかし，この規格は，
組織が，環境マネジメントシステムを他のマネジメ
ントシステムの要求事項に統合するために共通のア
プローチ及びリスクに基づく考え方を用いることが
できるようにしている．


　この規格は，適合を評価するために用いる要求事
項を規定している．組織は，次のいずれかの方法に
よって，この規格への適合を実証することができ
る．
―　自己決定し，自己宣言する．

―　適合について，組織に対して利害関係をもつ人
　　又はグループ，例えば顧客などによる確認を求
　　める．
―　自己宣言について組織外部の人又はグループに
　　よる確認を求める．
―　外部機関による環境マネジメントシステムの認
　　証・登録を求める．

Annex A provides explanatory information to prevent misinterpretation of the requirements of this International Standard. Annex B shows broad technical correspondence between the previous edition of this International Standard and this edition. Implementation guidance on environmental management systems is included in ISO 14004.

In this International Standard, the following verbal forms are used:

— "shall" indicates a requirement;

— "should" indicates a recommendation;

— "may" indicates a permission;
— "can" indicates a possibility or a capability.

Information marked as "NOTE" is intended to assist the understanding or use of the document. "Notes to entry" used in Clause 3 provide additional information that supplements the terminological data and can contain provisions relating to the

序文　43

附属書 A には，この規格の要求事項の誤った解釈を防ぐための説明を示す．附属書 B には，旧規格（**JIS Q 14001**:2004）とこの規格との間の広範な技術的対応を示す．環境マネジメントシステムの実施の手引は，**JIS Q 14004** に記載されている．

この規格では，次のような表現形式を用いている．

— "〜しなければならない"（shall）は，要求事項を示し，

— "〜することが望ましい"（should）は，推奨を示し，

— "〜してもよい"（may）は，許容を示し，

— "〜することができる"，"〜できる"，"〜し得る" など（can）は，可能性又は実現能力を示す．

"**注記**"に記載されている情報は，この規格の理解又は利用を助けるためのものである．箇条 3 で用いている "**注記**"は，用語データを補完する追加情報を示すほか，用語の使用に関する規定事項を含む場合もある．

use of a term.

The terms and definitions in Clause 3 are arranged in conceptual order, with an alphabetical index provided at the end of the document.

## 1   Scope

This International Standard specifies the requirements for an environmental management system that an organization can use to enhance its environmental performance. This International Standard is intended for use by an organization seeking to manage its environmental responsibilities in a systematic manner that contributes to the environmental pillar of sustainability.

This International Standard helps an organization achieve the intended outcomes of its environmental management system, which provide value for the environment, the organization itself and interested parties. Consistent with the organization's environmental policy, the intended outcomes of an environmental management system include:

—   enhancement of environmental performance;

箇条 3 の用語及び定義は，概念の順に配列し，巻末には五十音順及びアルファベット順の索引を記載した．

# 1 適用範囲

この規格は，組織が環境パフォーマンスを向上させるために用いることができる環境マネジメントシステムの要求事項について規定する．この規格は，持続可能性の"環境の柱"に寄与するような体系的な方法で組織の環境責任をマネジメントしようとする組織によって用いられることを意図している．

この規格は，組織が，環境，組織自体及び利害関係者に価値をもたらす環境マネジメントシステムの意図した成果を達成するために役立つ．環境マネジメントシステムの意図した成果は，組織の環境方針に整合して，次の事項を含む．

— 環境パフォーマンスの向上

## ISO 14001

— fulfilment of compliance obligations;

— achievement of environmental objectives.

This International Standard is applicable to any organization, regardless of size, type and nature, and applies to the environmental aspects of its activities, products and services that the organization determines it can either control or influence considering a life cycle perspective. This International Standard does not state specific environmental performance criteria.

This International Standard can be used in whole or in part to systematically improve environmental management. Claims of conformity to this International Standard, however, are not acceptable unless all its requirements are incorporated into an organization's environmental management system and fulfilled without exclusion.

## 1 適用範囲

47

— 順守義務を満たすこと
— 環境目標の達成

　この規格は，規模，業種・形態及び性質を問わず，どのような組織にも適用でき，組織がライフサイクルの視点を考慮して管理することができる又は影響を及ぼすことができると決定した，組織の活動，製品及びサービスの環境側面に適用する．この規格は，特定の環境パフォーマンス基準を規定するものではない．

　この規格は，環境マネジメントを体系的に改善するために，全体を又は部分的に用いることができる．しかし，この規格への適合の主張は，全ての要求事項が除外されることなく組織の環境マネジメントシステムに組み込まれ，満たされていない限り，容認されない．

　　**注記**　この規格の対応国際規格及びその対応の程度を表す記号を，次に示す．
ISO 14001:2015, Environmental management systems—Requirements with guidance for use (IDT)

## 2   Normative references

There are no normative references.

## 3   Terms and definitions

For the purposes of this document, the following terms and definitions apply.

## 3.1   Terms related to organization and leadership

**3.1.1**

**management system**

set of interrelated or interacting elements of an *organization* (3.1.4) to establish policies and *objectives* (3.2.5) and *processes* (3.3.5) to achieve those objectives

Note 1 to entry: A management system can address a single discipline or several disciplines (e.g. quality, environment, occupational health and safety, energy, financial management).

なお，対応の程度を表す記号 "IDT" は，**ISO/IEC Guide 21-1** に基づき，"一致している" ことを示す．

## 2 引用規格

この規格には，引用規格はない．

## 3 用語及び定義

この規格で用いる主な用語及び定義は，次による．

### 3.1 組織及びリーダーシップに関する用語

#### 3.1.1

**マネジメントシステム**（management system）

方針，**目的**（**3.2.5**）及びその目的を達成するための**プロセス**（**3.3.5**）を確立するための，相互に関連する又は相互に作用する，**組織**（**3.1.4**）の一連の要素．

> 注記 1　一つのマネジメントシステムは，単一又は複数の分野（例えば，品質マネジメント，環境マネジメント，労働安全衛生マネジメント，エネルギーマネジ

Note 2 to entry: The system elements include the organization's structure, roles and responsibilities, planning and operation, performance evaluation and improvement.

Note 3 to entry: The scope of a management system can include the whole of the organization, specific and identified functions of the organization, specific and identified sections of the organization, or one or more functions across a group of organizations.

**3.1.2**

**environmental management system**

part of the *management system* (3.1.1) used to manage *environmental aspects* (3.2.2), fulfil *compliance obligations* (3.2.9), and address *risks and opportunities* (3.2.11)

**3.1.3**

**environmental policy**

intentions and direction of an *organization* (3.1.4)

## 3 用語及び定義　　　51

メント，財務マネジメント）を取り扱うことができる．

**注記 2**　システムの要素には，組織の構造，役割及び責任，計画及び運用，パフォーマンス評価並びに改善が含まれる．

**注記 3**　マネジメントシステムの適用範囲としては，組織全体，組織内の固有で特定された機能，組織内の固有で特定された部門，複数の組織の集まりを横断する一つ又は複数の機能，などがあり得る．

**3.1.2**
**環境マネジメントシステム**（environmental management system）

　マネジメントシステム（**3.1.1**）の一部で，**環境側面**（**3.2.2**）をマネジメントし，**順守義務**（**3.2.9**）を満たし，**リスク及び機会**（**3.2.11**）に取り組むために用いられるもの．

**3.1.3**
**環境方針**（environmental policy）
　**トップマネジメント**（**3.1.5**）によって正式に表

related to *environmental performance* (3.4.11), as formally expressed by its *top management* (3.1.5)

**3.1.4**

**organization**

person or group of people that has its own functions with responsibilities, authorities and relationships to achieve its *objectives* (3.2.5)

Note 1 to entry: The concept of organization includes, but is not limited to sole-trader, company, corporation, firm, enterprise, authority, partnership, charity or institution, or part or combination thereof, whether incorporated or not, public or private.

**3.1.5**

**top management**

person or group of people who directs and controls an *organization* (3.1.4) at the highest level

Note 1 to entry: Top management has the power to delegate authority and provide resources within

明された，**環境パフォーマンス（3.4.11）**に関する，**組織（3.1.4）**の意図及び方向付け．

**3.1.4**
**組織**（organization）
　自らの**目的（3.2.5）**を達成するため，責任，権限及び相互関係を伴う独自の機能をもつ，個人又は人々の集まり．

　　　注記　組織という概念には，法人か否か，公的か私的かを問わず，自営業者，会社，法人，事務所，企業，当局，共同経営会社，非営利団体若しくは協会，又はこれらの一部若しくは組合せが含まれる．ただし，これらに限定されるものではない．

**3.1.5**
**トップマネジメント**（top management）
　最高位で**組織（3.1.4）**を指揮し，管理する個人又は人々の集まり．

　　　注記1　トップマネジメントは，組織内で，権限を委譲し，資源を提供する力をもっ

the organization.

Note 2 to entry: If the scope of the *management system* (3.1.1) covers only part of an organization, then top management refers to those who direct and control that part of the organization.

**3.1.6**

**interested party**

person or *organization* (3.1.4) that can affect, be affected by, or perceive itself to be affected by a decision or activity

EXAMPLE    Customers, communities, suppliers, regulators, non-governmental organizations, investors and employees.

Note 1 to entry: To "perceive itself to be affected" means the perception has been made known to the organization.

## 3.2   Terms related to planning

**3.2.1**

**environment**

surroundings in which an *organization* (3.1.4) op-

ている.

**注記2** マネジメントシステム（**3.1.1**）の適用範囲が組織の一部だけの場合，トップマネジメントとは，組織内のその一部を指揮し，管理する人をいう.

**3.1.6**
**利害関係者**（interested party）

ある決定事項若しくは活動に影響を与え得るか，その影響を受け得るか，又はその影響を受けると認識している，個人又は**組織**（**3.1.4**）.

例 顧客，コミュニティ，供給者，規制当局，非政府組織（NGO），投資家，従業員

注記 "影響を受けると認識している"とは，その認識が組織に知らされていることを意味している.

**3.2　計画に関する用語**
**3.2.1**
**環境**（environment）

大気，水，土地，天然資源，植物，動物，人及び

erates, including air, water, land, natural resources, flora, fauna, humans and their interrelationships

Note 1 to entry: Surroundings can extend from within an organization to the local, regional and global system.

Note 2 to entry: Surroundings can be described in terms of biodiversity, ecosystems, climate or other characteristics.

### 3.2.2

### environmental aspect

element of an *organization's* (3.1.4) activities or products or services that interacts or can interact with the *environment* (3.2.1)

Note 1 to entry: An environmental aspect can cause (an) *environmental impact(s)* (3.2.4). A significant environmental aspect is one that has or can have one or more significant environmental impact(s).

Note 2 to entry: Significant environmental aspects are determined by the organization applying one

それらの相互関係を含む，**組織（3.1.4）**の活動をとりまくもの．

注記 1　"とりまくもの"は，組織内から，近隣地域，地方及び地球規模のシステムにまで広がり得る．

注記 2　"とりまくもの"は，生物多様性，生態系，気候又はその他の特性の観点から表されることもある．

**3.2.2**

**環境側面**（environmental aspect）

**環境（3.2.1）**と相互に作用する，又は相互に作用する可能性のある，**組織（3.1.4）**の活動又は製品又はサービスの要素．

注記 1　環境側面は，**環境影響（3.2.4）**をもたらす可能性がある．著しい環境側面は，一つ又は複数の著しい環境影響を与える又は与える可能性がある．

注記 2　組織は，一つ又は複数の基準を適用して著しい環境側面を決定する．

58          ISO 14001

or more criteria.

### 3.2.3

### environmental condition

state or characteristic of the *environment* (3.2.1) as determined at a certain point in time

### 3.2.4

### environmental impact

change to the *environment* (3.2.1), whether adverse or beneficial, wholly or partially resulting from an *organization's* (3.1.4) *environmental aspects* (3.2.2)

### 3.2.5

### objective

result to be achieved

Note 1 to entry: An objective can be strategic, tactical, or operational.

Note 2 to entry: Objectives can relate to different disciplines (such as financial, health and safety, and environmental goals) and can apply at different levels (such as strategic, organization-wide, project, product, service and *process* (3.3.5)).

### 3.2.3

**環境状態**（environmental condition）

ある特定の時点において決定される，**環境**
（**3.2.1**）の様相又は特性．

### 3.2.4

**環境影響**（environmental impact）

有害か有益かを問わず，全体的に又は部分的に**組織**（**3.1.4**）の環境側面（**3.2.2**）から生じる，**環境**（**3.2.1**）に対する変化．

### 3.2.5

**目的，目標**（objective）

達成する結果．

> **注記 1**　目的（又は目標）は，戦略的，戦術的又は運用的であり得る．
>
> **注記 2**　目的（又は目標）は，様々な領域［例えば，財務，安全衛生，環境の到達点（goal）］に関連し得るものであり，様々な階層［例えば，戦略的レベル，組織全体，プロジェクト単位，製品ご

Note 3 to entry: An objective can be expressed in other ways, e.g. as an intended outcome, a purpose, an operational criterion, as an *environmental objective* (3.2.6), or by the use of other words with similar meaning (e.g. aim, goal, or target).

**3.2.6**

**environmental objective**

*objective* (3.2.5) set by the *organization* (3.1.4) consistent with its *environmental policy* (3.1.3)

**3.2.7**

**prevention of pollution**

use of *processes* (3.3.5), practices, techniques, materials, products, services or energy to avoid, reduce or control (separately or in combination) the creation, emission or discharge of any type of pollutant or waste, in order to reduce adverse *environmental impacts* (3.2.4)

と，サービスごと，**プロセス**（**3.3.5**）ごと］で適用できる．

注記 3　目的（又は目標）は，例えば，意図する成果，目的（purpose），運用基準など，別の形で表現することもできる．また，**環境目標**（**3.2.6**）という表現の仕方もある．又は，同じような意味をもつ別の言葉［**例**　狙い（aim），到達点（goal），目標（target）］で表すこともできる．

**3.2.6**

**環境目標**（environmental objective）

　**組織**（**3.1.4**）が設定する，**環境方針**（**3.1.3**）と整合のとれた**目標**（**3.2.5**）．

**3.2.7**

**汚染の予防**（prevention of pollution）

　有害な**環境影響**（**3.2.4**）を低減するために，様々な種類の汚染物質又は廃棄物の発生，排出又は放出を回避，低減又は管理するための**プロセス**（**3.3.5**），操作，技法，材料，製品，サービス又はエネルギーを（個別に又は組み合わせて）使用すること．

Note 1 to entry: Prevention of pollution can include source reduction or elimination; process, product or service changes; efficient use of resources; material and energy substitution; reuse; recovery; recycling, reclamation; or treatment.

**3.2.8**

**requirement**

need or expectation that is stated, generally implied or obligatory

Note 1 to entry: "Generally implied" means that it is custom or common practice for the *organization* (3.1.4) and *interested parties* (3.1.6) that the need or expectation under consideration is implied.

Note 2 to entry: A specified requirement is one that is stated, for example in *documented information* (3.3.2).

Note 3 to entry: Requirements other than legal requirements become obligatory when the organiza-

## 3 用語及び定義　　　63

注記　汚染の予防には，発生源の低減若しくは排除，プロセス，製品若しくはサービスの変更，資源の効率的な使用，代替材料及び代替エネルギーの利用，再利用，回収，リサイクル，再生又は処理が含まれ得る．

**3.2.8**
**要求事項**（requirement）

明示されている，通常暗黙のうちに了解されている又は義務として要求されている，ニーズ又は期待．

注記1　"通常暗黙のうちに了解されている"とは，対象となるニーズ又は期待が暗黙のうちに了解されていることが，**組織（3.1.4）**及び**利害関係者（3.1.6）**にとって，慣習又は慣行であることを意味する．

注記2　規定要求事項とは，例えば，**文書化した情報（3.3.2）**の中で明示されている要求事項をいう．

注記3　法的要求事項以外の要求事項は，組織がそれを順守することを決定したとき

tion decides to comply with them.

**3.2.9**

**compliance obligations** (preferred term)

legal requirements and other requirements (admitted term)

legal *requirements* (3.2.8) that an *organization* (3.1.4) has to comply with and other requirements that an organization has to or chooses to comply with

Note 1 to entry: Compliance obligations are related to the *environmental management system* (3.1.2).

Note 2 to entry: Compliance obligations can arise from mandatory requirements, such as applicable laws and regulations, or voluntary commitments, such as organizational and industry standards, contractual relationships, codes of practice and agreements with community groups or non-governmental organizations.

**3.2.10**

**risk**

effect of uncertainty

## 3　用語及び定義

に義務となる．

**3.2.9**

**順守義務**（compliance obligation）

**組織**（**3.1.4**）が順守しなければならない法的**要求事項**（**3.2.8**），及び組織が順守しなければならない又は順守することを選んだその他の要求事項．

注記 1　順守義務は，**環境マネジメントシステム**（**3.1.2**）に関連している．

注記 2　順守義務は，適用される法律及び規制のような強制的な要求事項から生じる場合もあれば，組織及び業界の標準，契約関係，行動規範，コミュニティグループ又は非政府組織（NGO）との合意のような，自発的なコミットメントから生じる場合もある．

**3.2.10**

**リスク**（risk）

不確かさの影響．

Note 1 to entry: An effect is a deviation from the expected — positive or negative.

Note 2 to entry: Uncertainty is the state, even partial, of deficiency of information related to, understanding or knowledge of, an event, its consequence, or likelihood.

Note 3 to entry: Risk is often characterized by reference to potential *"events"* (as defined in ISO Guide 73:2009, 3.5.1.3) and *"consequences"* (as defined in ISO Guide 73:2009, 3.6.1.3), or a combination of these.

Note 4 to entry: Risk is often expressed in terms of a combination of the consequences of an event (including changes in circumstances) and the associated *"likelihood"* (as defined in ISO Guide 73:2009, 3.6.1.1) of occurrence.

**3.2.11**

**risks and opportunities**

potential adverse effects (threats) and potential beneficial effects (opportunities)

### 3 用語及び定義　　　67

注記 1　影響とは，期待されていることから，好ましい方向又は好ましくない方向にかい（乖）離することをいう．

注記 2　不確かさとは，事象，その結果又はその起こりやすさに関する，情報，理解又は知識に，たとえ部分的にでも不備がある状態をいう．

注記 3　リスクは，起こり得る"事象"（**JIS Q 0073**:2010 の **3.5.1.3** の定義を参照．）及び"結果"（**JIS Q 0073**:2010 の **3.6.1.3** の定義を参照．），又はこれらの組合せについて述べることによって，その特徴を示すことが多い．

注記 4　リスクは，ある事象（その周辺状況の変化を含む．）の結果とその発生の"起こりやすさ"（**JIS Q 0073**:2010 の **3.6.1.1** の定義を参照．）との組合せとして表現されることが多い．

### 3.2.11
**リスク及び機会**（risks and opportunities）

潜在的で有害な影響（脅威）及び潜在的で有益な影響（機会）．

## 3.3 Terms related to support and operation

**3.3.1**

**competence**

ability to apply knowledge and skills to achieve intended results

**3.3.2**

**documented information**

information required to be controlled and maintained by an *organization* (3.1.4) and the medium on which it is contained

Note 1 to entry: Documented information can be in any format and media, and from any source.

Note 2 to entry: Documented information can refer to:

— the *environmental management system* (3.1.2), including related *processes* (3.3.5);

— information created in order for the organization to operate (can be referred to as documentation);

— evidence of results achieved (can be referred

## 3.3 支援及び運用に関する用語

### 3.3.1

**力量**（competence）

意図した結果を達成するために，知識及び技能を適用する能力．

### 3.3.2

**文書化した情報**（documented information）

組織（3.1.4）が管理し，維持するよう要求されている情報，及びそれが含まれている媒体．

注記 1　文書化した情報は，様々な形式及び媒体の形をとることができ，様々な情報源から得ることができる．

注記 2　文書化した情報には，次に示すものがあり得る．

— 関連する**プロセス**（3.3.5）を含む**環境マネジメントシステム**（3.1.2）

— 組織の運用のために作成された情報（文書類と呼ぶこともある．）

— 達成された結果の証拠（記録と呼

70           ISO 14001

to as records).

### 3.3.3
### life cycle

consecutive and interlinked stages of a product (or service) system, from raw material acquisition or generation from natural resources to final disposal

Note 1 to entry: The life cycle stages include acquisition of raw materials, design, production, transportation/delivery, use, end-of-life treatment and final disposal.

[SOURCE: ISO 14044:2006, 3.1, modified — The words "(or service)" have been added to the definition and Note 1 to entry has been added.]

### 3.3.4
### outsource (verb)

make an arrangement where an external *organization* (3.1.4) performs part of an organization's function or *process* (3.3.5)

Note 1 to entry: An external organization is out-

### 3　用語及び定義　　　　71

ぶこともある.)


**3.3.3**
**ライフサイクル**（life cycle）
　原材料の取得又は天然資源の産出から，最終処分までを含む，連続的でかつ相互に関連する製品（又はサービス）システムの段階群.

　　　注記　ライフサイクルの段階には，原材料の取得，設計，生産，輸送又は配送（提供），使用，使用後の処理及び最終処分が含まれる.


　［**JIS Q 14044**:2010 の **3.1** を変更."（又はサービス）" を追加し，文章構成を変更し，かつ，**注記**を追加している.］


**3.3.4**
**外部委託する**（outsource）（動詞）
　ある**組織**（**3.1.4**）の機能又は**プロセス**（**3.3.5**）の一部を外部の組織が実施するという取決めを行う.

　　　注記　外部委託した機能又はプロセスは**マネジ**

side the scope of the *management system* (3.1.1), although the outsourced function or process is within the scope.

**3.3.5**
**process**
set of interrelated or interacting activities which transforms inputs into outputs

Note 1 to entry: A process can be documented or not.

## 3.4 Terms related to performance evaluation and improvement
**3.4.1**
**audit**
systematic, independent and documented *process* (3.3.5) for obtaining audit evidence and evaluating it objectively to determine the extent to which the audit criteria are fulfilled

Note 1 to entry: An internal audit is conducted by the *organization* (3.1.4) itself, or by an external party on its behalf.

メントシステム（**3.1.1**）の適用範囲内
にあるが，外部の組織はマネジメントシ
ステムの適用範囲の外にある．

**3.3.5**
**プロセス**（process）
　インプットをアウトプットに変換する，相互に関
連する又は相互に作用する一連の活動．

　　注記　プロセスは，文書化することも，しない
　　　　　こともある．

## 3.4　パフォーマンス評価及び改善に関する用語

**3.4.1**
**監査**（audit）
　監査基準が満たされている程度を判定するため
に，監査証拠を収集し，それを客観的に評価する
ための，体系的で，独立し，文書化した**プロセス**
（**3.3.5**）．

　　注記1　内部監査は，その**組織**（**3.1.4**）自体
　　　　　　が行うか，又は組織の代理で外部関係
　　　　　　者が行う．

Note 2 to entry: An audit can be a combined audit (combining two or more disciplines).

Note 3 to entry: Independence can be demonstrated by the freedom from responsibility for the activity being audited or freedom from bias and conflict of interest.

Note 4 to entry: "Audit evidence" consists of records, statements of fact or other information which are relevant to the audit criteria and are verifiable; and "audit criteria" are the set of policies, procedures or *requirements* (3.2.8) used as a reference against which audit evidence is compared, as defined in ISO 19011:2011, 3.3 and 3.2 respectively.

**3.4.2**

**conformity**

fulfilment of a *requirement* (3.2.8)

**3.4.3**

**nonconformity**

non-fulfilment of a *requirement* (3.2.8)

Note 1 to entry: Nonconformity relates to require-

### 3 用語及び定義　　75

注記 2　監査は，複合監査（複数の分野の組合せ）でもあり得る．

注記 3　独立性は，監査の対象となる活動に関する責任を負っていないことで，又は偏り及び利害抵触がないことで，実証することができる．

注記 4　**JIS Q 19011**:2012 の **3.3** 及び **3.2** にそれぞれ定義されているように，"監査証拠"は，監査基準に関連し，かつ，検証できる，記録，事実の記述又はその他の情報から成り，"監査基準"は，監査証拠と比較する基準として用いる一連の方針，手順又は**要求事項**（**3.2.8**）である．


**3.4.2**

**適合**（conformity）

**要求事項**（**3.2.8**）を満たしていること．


**3.4.3**

**不適合**（nonconformity）

**要求事項**（**3.2.8**）を満たしていないこと．

注記　不適合は，この規格に規定する要求事

ments in this International Standard and additional *environmental management system* (3.1.2) requirements that an *organization* (3.1.4) establishes for itself.

### 3.4.4
### corrective action

action to eliminate the cause of a *nonconformity* (3.4.3) and to prevent recurrence

Note 1 to entry: There can be more than one cause for a nonconformity.

### 3.4.5
### continual improvement

recurring activity to enhance *performance* (3.4.10)

Note 1 to entry: Enhancing performance relates to the use of the *environmental management system* (3.1.2) to enhance *environmental performance* (3.4.11) consistent with the *organization's* (3.1.4) *environmental policy* (3.1.3).

項，及び**組織（3.1.4）**が自ら定める追加的な**環境マネジメントシステム（3.1.2）**要求事項に関連している．

**3.4.4**

**是正処置**（corrective action）

**不適合（3.4.3）**の原因を除去し，再発を防止するための処置．

　　　注記　不適合には，複数の原因がある場合がある．

**3.4.5**

**継続的改善**（continual improvement）

**パフォーマンス（3.4.10）**を向上するために繰り返し行われる活動．

　　　注記1　パフォーマンスの向上は，**組織（3.1.4）**の**環境方針（3.1.3）**と整合して**環境パフォーマンス（3.4.11）**を向上するために，**環境マネジメントシステム（3.1.2）**を用いることに関連している．

Note 2 to entry: The activity need not take place in all areas simultaneously, or without interruption.

**3.4.6**
**effectiveness**
extent to which planned activities are realized and planned results achieved

**3.4.7**
**indicator**
measurable representation of the condition or status of operations, management or conditions

[SOURCE: ISO 14031:2013, 3.15]

**3.4.8**
**monitoring**
determining the status of a system, a *process* (3.3.5) or an activity

Note 1 to entry: To determine the status, there might be a need to check, supervise or critically observe.

注記2　活動は，必ずしも全ての領域で同時
に，又は中断なく行う必要はない.

## 3.4.6
**有効性**（effectiveness）
　計画した活動を実行し，計画した結果を達成した
程度.

## 3.4.7
**指標**（indicator）
　運用，マネジメント又は条件の状態又は状況の，
測定可能な表現.

（**ISO 14031**:2013 の **3.15** 参照）

## 3.4.8
**監視**（monitoring）
　システム，**プロセス**（**3.3.5**）又は活動の状況を
明確にすること.

注記　状況を明確にするために，点検，監督又
は注意深い観察が必要な場合もある.

**3.4.9**

**measurement**

*process* (3.3.5) to determine a value

**3.4.10**

**performance**

measurable result

Note 1 to entry: Performance can relate either to quantitative or qualitative findings.

Note 2 to entry: Performance can relate to the management of activities, *processes* (3.3.5), products (including services), systems or *organizations* (3.1.4).

**3.4.11**

**environmental performance**

*performance* (3.4.10) related to the management of *environmental aspects* (3.2.2)

Note 1 to entry: For an *environmental management system* (3.1.2), results can be measured against the *organization's* (3.1.4) *environmental policy* (3.1.3), *environmental objectives* (3.2.6) or other criteria,

### 3　用語及び定義

**3.4.9**
**測定**（measurement）
　値を決定する**プロセス**（**3.3.5**）．

**3.4.10**
**パフォーマンス**（performance）
　測定可能な結果．

> 注記 1　パフォーマンスは，定量的又は定性的
> 　　　　な所見のいずれにも関連し得る．
> 注記 2　パフォーマンスは，活動，**プロセス**
> 　　　　（**3.3.5**），製品（サービスを含む．），
> 　　　　システム又は**組織**（**3.1.4**）の運営管
> 　　　　理に関連し得る．

**3.4.11**
**環境パフォーマンス**（environmental performance）
　**環境側面**（**3.2.2**）のマネジメントに関連する**パ
フォーマンス**（**3.4.10**）．

> 注記　環境マネジメントシステム（**3.1.2**）で
> 　　　は，結果は，**組織**（**3.1.4**）の**環境方針**
> 　　　（**3.1.3**），**環境目標**（**3.2.6**），又はその
> 　　　他の基準に対して，**指標**（**3.4.7**）を用

using *indicators* (3.4.7).

# 4 Context of the organization

## 4.1 Understanding the organization and its context

The organization shall determine external and internal issues that are relevant to its purpose and that affect its ability to achieve the intended outcomes of its environmental management system. Such issues shall include environmental conditions being affected by or capable of affecting the organization.

## 4.2 Understanding the needs and expectations of interested parties

The organization shall determine:

a)  the interested parties that are relevant to the environmental management system;

b)  the relevant needs and expectations (i.e. requirements) of these interested parties;

c)  which of these needs and expectations become its compliance obligations.

いて測定可能である．

## 4 組織の状況
### 4.1 組織及びその状況の理解

組織は，組織の目的に関連し，かつ，その環境マネジメントシステムの意図した成果を達成する組織の能力に影響を与える，外部及び内部の課題を決定しなければならない．こうした課題には，組織から影響を受ける又は組織に影響を与える可能性がある環境状態を含めなければならない．

### 4.2 利害関係者のニーズ及び期待の理解

組織は，次の事項を決定しなければならない．

a) 環境マネジメントシステムに関連する利害関係者

b) それらの利害関係者の，関連するニーズ及び期待（すなわち，要求事項）

c) それらのニーズ及び期待のうち，組織の順守義務となるもの

## 4.3 Determining the scope of the environmental management system

The organization shall determine the boundaries and applicability of the environmental management system to establish its scope.

When determining this scope, the organization shall consider:

a)  the external and internal issues referred to in 4.1;

b)  the compliance obligations referred to in 4.2;

c)  its organizational units, functions and physical boundaries;

d)  its activities, products and services;

e)  its authority and ability to exercise control and influence.

Once the scope is defined, all activities, products and services of the organization within that scope need to be included in the environmental management system.

The scope shall be maintained as documented information and be available to interested parties.

## 4.3 環境マネジメントシステムの適用範囲の決定

組織は，環境マネジメントシステムの適用範囲を定めるために，その境界及び適用可能性を決定しなければならない．

この適用範囲を決定するとき，組織は，次の事項を考慮しなければならない．

a) 4.1 に規定する外部及び内部の課題

b) 4.2 に規定する順守義務

c) 組織の単位，機能及び物理的境界

d) 組織の活動，製品及びサービス

e) 管理し影響を及ぼす，組織の権限及び能力

適用範囲が定まれば，その適用範囲の中にある組織の全ての活動，製品及びサービスは，環境マネジメントシステムに含まれている必要がある．

環境マネジメントシステムの適用範囲は，文書化した情報として維持しなければならず，かつ，利害

## 4.4 Environmental management system

To achieve the intended outcomes, including enhancing its environmental performance, the organization shall establish, implement, maintain and continually improve an environmental management system, including the processes needed and their interactions, in accordance with the requirements of this International Standard.

The organization shall consider the knowledge gained in 4.1 and 4.2 when establishing and maintaining the environmental management system.

## 5 Leadership

## 5.1 Leadership and commitment

Top management shall demonstrate leadership and commitment with respect to the environmental management system by:

a)  taking accountability for the effectiveness of the environmental management system;

b)  ensuring that the environmental policy and

関係者がこれを入手できるようにしなければならない.

## 4.4　環境マネジメントシステム

環境パフォーマンスの向上を含む意図した成果を達成するため,組織は,この規格の要求事項に従って,必要なプロセス及びそれらの相互作用を含む,環境マネジメントシステムを確立し,実施し,維持し,かつ,継続的に改善しなければならない.

環境マネジメントシステムを確立し維持するとき,組織は,**4.1** 及び **4.2** で得た知識を考慮しなければならない.

# 5　リーダーシップ
## 5.1　リーダーシップ及びコミットメント

トップマネジメントは,次に示す事項によって,環境マネジメントシステムに関するリーダーシップ及びコミットメントを実証しなければならない.

**a)**　環境マネジメントシステムの有効性に説明責任を負う.

**b)**　環境方針及び環境目標を確立し,それらが組織

environmental objectives are established and are compatible with the strategic direction and the context of the organization;

c)  ensuring the integration of the environmental management system requirements into the organization's business processes;

d)  ensuring that the resources needed for the environmental management system are available;

e)  communicating the importance of effective environmental management and of conforming to the environmental management system requirements;

f)  ensuring that the environmental management system achieves its intended outcomes;

g)  directing and supporting persons to contribute to the effectiveness of the environmental management system;

h)  promoting continual improvement;

i)  supporting other relevant management roles to demonstrate their leadership as it applies to their areas of responsibility.

NOTE    Reference to "business" in this Inter-

## 5 リーダーシップ 89

の戦略的な方向性及び組織の状況と両立すること を確実にする.

c) 組織の事業プロセスへの環境マネジメントシステム要求事項の統合を確実にする.

d) 環境マネジメントシステムに必要な資源が利用可能であることを確実にする.

e) 有効な環境マネジメント及び環境マネジメントシステム要求事項への適合の重要性を伝達する.

f) 環境マネジメントシステムがその意図した成果を達成することを確実にする.

g) 環境マネジメントシステムの有効性に寄与するよう人々を指揮し,支援する.

h) 継続的改善を促進する.

i) その他の関連する管理層がその責任の領域においてリーダーシップを実証するよう,管理層の役割を支援する.

**注記** この規格で"事業"という場合,それ

national Standard can be interpreted broadly to mean those activities that are core to the purposes of the organization's existence.

## 5.2 Environmental policy

Top management shall establish, implement and maintain an environmental policy that, within the defined scope of its environmental management system:

a) is appropriate to the purpose and context of the organization, including the nature, scale and environmental impacts of its activities, products and services;

b) provides a framework for setting environmental objectives;

c) includes a commitment to the protection of the environment, including prevention of pollution and other specific commitment(s) relevant to the context of the organization;

NOTE    Other specific commitment(s) to protect the environment can include sustainable resource use, climate change mitigation and adaptation, and protection of biodiversity and

は，組織の存在の目的の中核となる活動という広義の意味で解釈され得る．

## 5.2 環境方針

トップマネジメントは，組織の環境マネジメントシステムの定められた適用範囲の中で，次の事項を満たす環境方針を確立し，実施し，維持しなければならない．

**a)** 組織の目的，並びに組織の活動，製品及びサービスの性質，規模及び環境影響を含む組織の状況に対して適切である．

**b)** 環境目標の設定のための枠組みを示す．

**c)** 汚染の予防，及び組織の状況に関連するその他の固有なコミットメントを含む，環境保護に対するコミットメントを含む．

　　　**注記** 環境保護に対するその他の固有なコミットメントには，持続可能な資源の利用，気候変動の緩和及び気候変動への適応，並びに生物多様性及び生態系の

92          ISO 14001

ecosystems.

d)    includes a commitment to fulfil its compliance obligations;

e)    includes a commitment to continual improvement of the environmental management system to enhance environmental performance.

The environmental policy shall:

—    be maintained as documented information;

—    be communicated within the organization;

—    be available to interested parties.

## 5.3   Organizational roles, responsibilities and authorities

Top management shall ensure that the responsibilities and authorities for relevant roles are assigned and communicated within the organization.

Top management shall assign the responsibility and authority for:

a)    ensuring that the environmental management system conforms to the requirements of

5 リーダーシップ　　　93

保護を含み得る.

**d)** 組織の順守義務を満たすことへのコミットメントを含む.

**e)** 環境パフォーマンスを向上させるための環境マネジメントシステムの継続的改善へのコミットメントを含む.

環境方針は,次に示す事項を満たさなければならない.
— 文書化した情報として維持する.
— 組織内に伝達する.
— 利害関係者が入手可能である.

## 5.3　組織の役割,責任及び権限

トップマネジメントは,関連する役割に対して,責任及び権限が割り当てられ,組織内に伝達されることを確実にしなければならない.

トップマネジメントは,次の事項に対して,責任及び権限を割り当てなければならない.
**a)** 環境マネジメントシステムが,この規格の要求事項に適合することを確実にする.

this International Standard;

b)  reporting on the performance of the environmental management system, including environmental performance, to top management.

# 6  Planning

## 6.1  Actions to address risks and opportunities

### 6.1.1  General

The organization shall establish, implement and maintain the process(es) needed to meet the requirements in 6.1.1 to 6.1.4.

When planning for the environmental management system, the organization shall consider:

a)  the issues referred to in 4.1;

b)  the requirements referred to in 4.2;

c)  the scope of its environmental management system;

and determine the risks and opportunities, related to its environmental aspects (see 6.1.2), compliance obligations (see 6.1.3) and other issues and requirements, identified in 4.1 and 4.2, that need

**b)** 環境パフォーマンスを含む環境マネジメントシステムのパフォーマンスをトップマネジメントに報告する.

## 6 計画
### 6.1 リスク及び機会への取組み

#### 6.1.1 一般
組織は, **6.1.1〜6.1.4** に規定する要求事項を満たすために必要なプロセスを確立し, 実施し, 維持しなければならない.

環境マネジメントシステムの計画を策定するとき, 組織は, 次の **a)〜c)** を考慮し,
**a)** **4.1** に規定する課題
**b)** **4.2** に規定する要求事項
**c)** 環境マネジメントシステムの適用範囲

次の事項のために取り組む必要がある, 環境側面 (**6.1.2** 参照), 順守義務 (**6.1.3** 参照), 並びに **4.1** 及び **4.2** で特定したその他の課題及び要求事項に関連する, リスク及び機会を決定しなければならな

to be addressed to:

— give assurance that the environmental management system can achieve its intended outcomes;
— prevent or reduce undesired effects, including the potential for external environmental conditions to affect the organization;
— achieve continual improvement.

Within the scope of the environmental management system, the organization shall determine potential emergency situations, including those that can have an environmental impact.

The organization shall maintain documented information of its:

— risks and opportunities that need to be addressed;
— process(es) needed in 6.1.1 to 6.1.4, to the extent necessary to have confidence they are carried out as planned.

### 6.1.2   Environmental aspects

Within the defined scope of the environmental

い.
— 環境マネジメントシステムが，その意図した成果を達成できるという確信を与える.

— 外部の環境状態が組織に影響を与える可能性を含め，望ましくない影響を防止又は低減する.

— 継続的改善を達成する.

組織は，環境マネジメントシステムの適用範囲の中で，環境影響を与える可能性のあるものを含め，潜在的な緊急事態を決定しなければならない.

組織は，次に関する文書化した情報を維持しなければならない.
— 取り組む必要があるリスク及び機会

— **6.1.1～6.1.4** で必要なプロセスが計画どおりに実施されるという確信をもつために必要な程度の，それらのプロセス

## 6.1.2 環境側面
組織は，環境マネジメントシステムの定められた

management system, the organization shall determine the environmental aspects of its activities, products and services that it can control and those that it can influence, and their associated environmental impacts, considering a life cycle perspective.

When determining environmental aspects, the organization shall take into account:

a)  change, including planned or new developments, and new or modified activities, products and services;

b)  abnormal conditions and reasonably foreseeable emergency situations.

The organization shall determine those aspects that have or can have a significant environmental impact, i.e. significant environmental aspects, by using established criteria.

The organization shall communicate its significant environmental aspects among the various levels and functions of the organization, as appropriate.

## 6 計 画

適用範囲の中で，ライフサイクルの視点を考慮し，組織の活動，製品及びサービスについて，組織が管理できる環境側面及び組織が影響を及ぼすことができる環境側面，並びにそれらに伴う環境影響を決定しなければならない．

環境側面を決定するとき，組織は，次の事項を考慮に入れなければならない．

**a)** 変更．これには，計画した又は新規の開発，並びに新規の又は変更された活動，製品及びサービスを含む．

**b)** 非通常の状況及び合理的に予見できる緊急事態

組織は，設定した基準を用いて，著しい環境影響を与える又は与える可能性のある側面（すなわち，著しい環境側面）を決定しなければならない．

組織は，必要に応じて，組織の種々の階層及び機能において，著しい環境側面を伝達しなければならない．

The organization shall maintain documented information of its:

— environmental aspects and associated environmental impacts;

— criteria used to determine its significant environmental aspects;

— significant environmental aspects.

NOTE Significant environmental aspects can result in risks and opportunities associated with either adverse environmental impacts (threats) or beneficial environmental impacts (opportunities).

### 6.1.3 Compliance obligations

The organization shall:

a) determine and have access to the compliance obligations related to its environmental aspects;

b) determine how these compliance obligations apply to the organization;

c) take these compliance obligations into account when establishing, implementing, maintaining and continually improving its environmental management system.

6　計　　画　　　　101

　組織は，次に関する文書化した情報を維持しなければならない．
— 　環境側面及びそれに伴う環境影響

— 　著しい環境側面を決定するために用いた基準

— 　著しい環境側面

　　注記　著しい環境側面は，有害な環境影響（脅威）又は有益な環境影響（機会）に関連するリスク及び機会をもたらし得る．


### 6.1.3　順守義務

　組織は，次の事項を行わなければならない．

a)　組織の環境側面に関する順守義務を決定し，参照する．

b)　これらの順守義務を組織にどのように適用するかを決定する．

c)　環境マネジメントシステムを確立し，実施し，維持し，継続的に改善するときに，これらの順守義務を考慮に入れる．

102　　　　　　　　ISO 14001

The organization shall maintain documented information of its compliance obligations.

NOTE　　Compliance obligations can result in risks and opportunities to the organization.

### 6.1.4　Planning action

The organization shall plan:

a)　to take actions to address its:

　　1)　significant environmental aspects;

　　2)　compliance obligations;

　　3)　risks and opportunities identified in 6.1.1;

b)　how to:

　　1)　integrate and implement the actions into its environmental management system processes (see 6.2, Clause 7, Clause 8 and 9.1), or other business processes;

　　2)　evaluate the effectiveness of these actions (see 9.1).

When planning these actions, the organization shall consider its technological options and its financial, operational and business requirements.

## 6　計　画

組織は，順守義務に関する文書化した情報を維持しなければならない．

> **注記**　順守義務は，組織に対するリスク及び機会をもたらし得る．

### 6.1.4　取組みの計画策定

組織は，次の事項を計画しなければならない．

**a)**　次の事項への取組み

**1)**　著しい環境側面

**2)**　順守義務

**3)**　**6.1.1** で特定したリスク及び機会

**b)**　次の事項を行う方法

**1)**　その取組みの環境マネジメントシステムプロセス（**6.2**，箇条 **7**，箇条 **8** 及び **9.1** 参照）又は他の事業プロセスへの統合及び実施

**2)**　その取組みの有効性の評価（**9.1** 参照）

これらの取組みを計画するとき，組織は，技術上の選択肢，並びに財務上，運用上及び事業上の要求事項を考慮しなければならない．

## 6.2 Environmental objectives and planning to achieve them

### 6.2.1 Environmental objectives

The organization shall establish environmental objectives at relevant functions and levels, taking into account the organization's significant environmental aspects and associated compliance obligations, and considering its risks and opportunities.

The environmental objectives shall be:

a) consistent with the environmental policy;

b) measurable (if practicable);

c) monitored;

d) communicated;

e) updated as appropriate.

The organization shall maintain documented information on the environmental objectives.

### 6.2.2 Planning actions to achieve environmental objectives

When planning how to achieve its environmental objectives, the organization shall determine:

## 6.2 環境目標及びそれを達成するための計画策定

### 6.2.1 環境目標

組織は，組織の著しい環境側面及び関連する順守義務を考慮に入れ，かつ，リスク及び機会を考慮し，関連する機能及び階層において，環境目標を確立しなければならない．

環境目標は，次の事項を満たさなければならない．

a) 環境方針と整合している．

b) （実行可能な場合）測定可能である．

c) 監視する．

d) 伝達する．

e) 必要に応じて，更新する．

組織は，環境目標に関する文書化した情報を維持しなければならない．

### 6.2.2 環境目標を達成するための取組みの計画策定

組織は，環境目標をどのように達成するかについて計画するとき，次の事項を決定しなければならな

106                    ISO 14001

a)    what will be done;

b)    what resources will be required;

c)    who will be responsible;

d)    when it will be completed;

e)    how the results will be evaluated, including
      indicators for monitoring progress toward
      achievement of its measurable environmental
      objectives (see 9.1.1).


The organization shall consider how actions to
achieve its environmental objectives can be integrated into the organization's business processes.


## 7   Support
### 7.1   Resources
The organization shall determine and provide the
resources needed for the establishment, implementation, maintenance and continual improvement of
the environmental management system.


### 7.2   Competence
The organization shall:

a)    determine    the    necessary    competence    of

い.

a) 実施事項

b) 必要な資源

c) 責任者

d) 達成期限

e) 結果の評価方法．これには，測定可能な環境目標の達成に向けた進捗を監視するための指標を含む（**9.1.1** 参照）．


組織は，環境目標を達成するための取組みを組織の事業プロセスにどのように統合するかについて，考慮しなければならない．

# 7 支援

## 7.1 資源

組織は，環境マネジメントシステムの確立，実施，維持及び継続的改善に必要な資源を決定し，提供しなければならない．


## 7.2 力量

組織は，次の事項を行わなければならない．

a) 組織の環境パフォーマンスに影響を与える業

person(s) doing work under its control that affects its environmental performance and its ability to fulfil its compliance obligations;

b) ensure that these persons are competent on the basis of appropriate education, training or experience;

c) determine training needs associated with its environmental aspects and its environmental management system;

d) where applicable, take actions to acquire the necessary competence, and evaluate the effectiveness of the actions taken.

NOTE Applicable actions can include, for example, the provision of training to, the mentoring of, or the reassignment of currently employed persons; or the hiring or contracting of competent persons.

The organization shall retain appropriate documented information as evidence of competence.

## 7.3   Awareness

務，及び順守義務を満たす組織の能力に影響を与える業務を組織の管理下で行う人（又は人々）に必要な力量を決定する．

b) 適切な教育，訓練又は経験に基づいて，それらの人々が力量を備えていることを確実にする．

c) 組織の環境側面及び環境マネジメントシステムに関する教育訓練のニーズを決定する．

d) 該当する場合には，必ず，必要な力量を身に付けるための処置をとり，とった処置の有効性を評価する．

注記 適用される処置には，例えば，現在雇用している人々に対する，教育訓練の提供，指導の実施，配置転換の実施などがあり，また，力量を備えた人々の雇用，そうした人々との契約締結などもあり得る．

組織は，力量の証拠として，適切な文書化した情報を保持しなければならない．

**7.3 認識**

The organization shall ensure that persons doing work under the organization's control are aware of:

a)   the environmental policy;

b)   the significant environmental aspects and related actual or potential environmental impacts associated with their work;

c)   their contribution to the effectiveness of the environmental management system, including the benefits of enhanced environmental performance;

d)   the implications of not conforming with the environmental management system requirements, including not fulfilling the organization's compliance obligations.


## 7.4   Communication

### 7.4.1   General

The organization shall establish, implement and maintain the process(es) needed for internal and external communications relevant to the environmental management system, including:

a)   on what it will communicate;

b)   when to communicate;

c)   with whom to communicate;

### 7 支援 111

組織は，組織の管理下で働く人々が次の事項に関して認識をもつことを確実にしなければならない．

**a)** 環境方針

**b)** 自分の業務に関係する著しい環境側面及びそれに伴う顕在する又は潜在的な環境影響

**c)** 環境パフォーマンスの向上によって得られる便益を含む，環境マネジメントシステムの有効性に対する自らの貢献

**d)** 組織の順守義務を満たさないことを含む，環境マネジメントシステム要求事項に適合しないことの意味

## 7.4 コミュニケーション

### 7.4.1 一般

組織は，次の事項を含む，環境マネジメントシステムに関連する内部及び外部のコミュニケーションに必要なプロセスを確立し，実施し，維持しなければならない．

**a)** コミュニケーションの内容

**b)** コミュニケーションの実施時期

**c)** コミュニケーションの対象者

d)   how to communicate.

When establishing its communication process(es), the organization shall:

— take into account its compliance obligations;

— ensure that environmental information communicated is consistent with information generated within the environmental management system, and is reliable.

The organization shall respond to relevant communications on its environmental management system.

The organization shall retain documented information as evidence of its communications, as appropriate.

### 7.4.2   Internal communication

The organization shall:

a)   internally communicate information relevant to the environmental management system among the various levels and functions of the organization, including changes to the envi-

## 7 支援　　　　113

**d)** コミュニケーションの方法

　コミュニケーションプロセスを確立するとき，組織は，次の事項を行わなければならない．
— 順守義務を考慮に入れる．
— 伝達される環境情報が，環境マネジメントシステムにおいて作成される情報と整合し，信頼性があることを確実にする．

　組織は，環境マネジメントシステムについての関連するコミュニケーションに対応しなければならない．

　組織は，必要に応じて，コミュニケーションの証拠として，文書化した情報を保持しなければならない．

**7.4.2　内部コミュニケーション**
　組織は，次の事項を行わなければならない．
**a)**　必要に応じて，環境マネジメントシステムの変更を含め，環境マネジメントシステムに関連する情報について，組織の種々の階層及び機能間で内部コミュニケーションを行う．

ronmental management system, as appropriate;

b) ensure its communication process(es) enable(s) persons doing work under the organization's control to contribute to continual improvement.

### 7.4.3 External communication

The organization shall externally communicate information relevant to the environmental management system, as established by the organization's communication process(es) and as required by its compliance obligations.

### 7.5 Documented information
### 7.5.1 General

The organization's environmental management system shall include:

a) documented information required by this International Standard;

b) documented information determined by the organization as being necessary for the effectiveness of the environmental management system.

**b)** コミュニケーションプロセスが，組織の管理下で働く人々の継続的改善への寄与を可能にすることを確実にする．

### 7.4.3 外部コミュニケーション

組織は，コミュニケーションプロセスによって確立したとおりに，かつ，順守義務による要求に従って，環境マネジメントシステムに関連する情報について外部コミュニケーションを行わなければならない．

### 7.5 文書化した情報
### 7.5.1 一般

組織の環境マネジメントシステムは，次の事項を含まなければならない．

**a)** この規格が要求する文書化した情報

**b)** 環境マネジメントシステムの有効性のために必要であると組織が決定した，文書化した情報

NOTE    The extent of documented information for an environmental management system can differ from one organization to another due to:

— the size of organization and its type of activities, processes, products and services;
— the need to demonstrate fulfilment of its compliance obligations;
— the complexity of processes and their interactions;
— the competence of persons doing work under the organization's control.

## 7.5.2  Creating and updating

When creating and updating documented information, the organization shall ensure appropriate:

a)  identification and description (e.g. a title, date, author, or reference number);

b)  format (e.g. language, software version, graphics) and media (e.g. paper, electronic);

c)  review and approval for suitability and adequacy.

## 7.5.3  Control of documented information

7 支 援　　117

注記　環境マネジメントシステムのための文書
　　　化した情報の程度は，次のような理由に
　　　よって，それぞれの組織で異なる場合が
　　　ある．
　　─　組織の規模，並びに活動，プロセ
　　　　ス，製品及びサービスの種類
　　─　順守義務を満たしていることを実証
　　　　する必要性
　　─　プロセス及びその相互作用の複雑さ

　　─　組織の管理下で働く人々の力量


## 7.5.2　作成及び更新

　文書化した情報を作成及び更新する際，組織は，
次の事項を確実にしなければならない．

a)　適切な識別及び記述（例えば，タイトル，日
　　付，作成者，参照番号）

b)　適切な形式（例えば，言語，ソフトウェアの
　　版，図表）及び媒体（例えば，紙，電子媒体）

c)　適切性及び妥当性に関する，適切なレビュー及
　　び承認


## 7.5.3　文書化した情報の管理

Documented information required by the environmental management system and by this International Standard shall be controlled to ensure:

a) it is available and suitable for use, where and when it is needed;

b) it is adequately protected (e.g. from loss of confidentiality, improper use, or loss of integrity).

For the control of documented information, the organization shall address the following activities as applicable:

— distribution, access, retrieval and use;

— storage and preservation, including preservation of legibility;

— control of changes (e.g. version control);

— retention and disposition.

Documented information of external origin determined by the organization to be necessary for the planning and operation of the environmental management system shall be identified, as appropriate, and controlled.

## 7 支　援

　環境マネジメントシステム及びこの規格で要求されている文書化した情報は，次の事項を確実にするために，管理しなければならない.

**a)** 文書化した情報が，必要なときに，必要なところで，入手可能かつ利用に適した状態である.

**b)** 文書化した情報が十分に保護されている（例えば，機密性の喪失，不適切な使用及び完全性の喪失からの保護）.


　文書化した情報の管理に当たって，組織は，該当する場合には，必ず，次の行動に取り組まなければならない.

— 配付，アクセス，検索及び利用

— 読みやすさが保たれることを含む，保管及び保存

— 変更の管理（例えば，版の管理）

— 保持及び廃棄


　環境マネジメントシステムの計画及び運用のために組織が必要と決定した外部からの文書化した情報は，必要に応じて識別し，管理しなければならない.

NOTE    Access can imply a decision regarding the permission to view the documented information only, or the permission and authority to view and change the documented information.

# 8    Operation

## 8.1    Operational planning and control

The organization shall establish, implement, control and maintain the processes needed to meet environmental management system requirements, and to implement the actions identified in 6.1 and 6.2, by:

— establishing operating criteria for the process(es);

— implementing control of the process(es), in accordance with the operating criteria.

NOTE    Controls can include engineering controls and procedures. Controls can be implemented following a hierarchy (e.g. elimination, substitution, administrative) and can be used individually or in combination.

The organization shall control planned changes

注記 アクセスとは，文書化した情報の閲覧だけの許可に関する決定，又は文書化した情報の閲覧及び変更の許可及び権限に関する決定を意味し得る．

# 8 運用

## 8.1 運用の計画及び管理

組織は，次に示す事項の実施によって，環境マネジメントシステム要求事項を満たすため，並びに**6.1**及び**6.2**で特定した取組みを実施するために必要なプロセスを確立し，実施し，管理し，かつ，維持しなければならない．

— プロセスに関する運用基準の設定

— その運用基準に従った，プロセスの管理の実施

注記 管理は，工学的な管理及び手順を含み得る．管理は，優先順位（例えば，除去，代替，管理的な対策）に従って実施されることもあり，また，個別に又は組み合わせて用いられることもある．

組織は，計画した変更を管理し，意図しない変更

## 122 ISO 14001

and review the consequences of unintended changes, taking action to mitigate any adverse effects, as necessary.

The organization shall ensure that outsourced processes are controlled or influenced. The type and extent of control or influence to be applied to the process(es) shall be defined within the environmental management system.

Consistent with a life cycle perspective, the organization shall:

a) establish controls, as appropriate, to ensure that its environmental requirement(s) is (are) addressed in the design and development process for the product or service, considering each life cycle stage;

b) determine its environmental requirement(s) for the procurement of products and services, as appropriate;

c) communicate its relevant environmental requirement(s) to external providers, including contractors;

d) consider the need to provide information about

によって生じた結果をレビューし，必要に応じて，有害な影響を緩和する処置をとらなければならない．

組織は，外部委託したプロセスが管理されている又は影響を及ぼされていることを確実にしなければならない．これらのプロセスに適用される，管理する又は影響を及ぼす方式及び程度は，環境マネジメントシステムの中で定めなければならない．

ライフサイクルの視点に従って，組織は，次の事項を行わなければならない．

**a)** 必要に応じて，ライフサイクルの各段階を考慮して，製品又はサービスの設計及び開発プロセスにおいて，環境上の要求事項が取り組まれていることを確実にするために，管理を確立する．

**b)** 必要に応じて，製品及びサービスの調達に関する環境上の要求事項を決定する．

**c)** 請負者を含む外部提供者に対して，関連する環境上の要求事項を伝達する．

**d)** 製品及びサービスの輸送又は配送（提供），使

potential significant environmental impacts associated with the transportation or delivery, use, end-of-life treatment and final disposal of its products and services.

The organization shall maintain documented information to the extent necessary to have confidence that the processes have been carried out as planned.

## 8.2 Emergency preparedness and response

The organization shall establish, implement and maintain the process(es) needed to prepare for and respond to potential emergency situations identified in 6.1.1.

The organization shall:

a) prepare to respond by planning actions to prevent or mitigate adverse environmental impacts from emergency situations;

b) respond to actual emergency situations;

c) take action to prevent or mitigate the consequences of emergency situations, appropriate to the magnitude of the emergency and the po-

用，使用後の処理及び最終処分に伴う潜在的な
著しい環境影響に関する情報を提供する必要性
について考慮する．

組織は，プロセスが計画どおりに実施されたとい
う確信をもつために必要な程度の，文書化した情報
を維持しなければならない．

## 8.2　緊急事態への準備及び対応

組織は，**6.1.1** で特定した潜在的な緊急事態への
準備及び対応のために必要なプロセスを確立し，実
施し，維持しなければならない．

組織は，次の事項を行わなければならない．

**a)**　緊急事態からの有害な環境影響を防止又は緩和
するための処置を計画することによって，対応
を準備する．

**b)**　顕在した緊急事態に対応する．

**c)**　緊急事態及びその潜在的な環境影響の大きさに
応じて，緊急事態による結果を防止又は緩和す
るための処置をとる．

tential environmental impact;

d) periodically test the planned response actions, where practicable;

e) periodically review and revise the process(es) and planned response actions, in particular after the occurrence of emergency situations or tests;

f) provide relevant information and training related to emergency preparedness and response, as appropriate, to relevant interested parties, including persons working under its control.

The organization shall maintain documented information to the extent necessary to have confidence that the process(es) is (are) carried out as planned.

# 9 Performance evaluation

## 9.1 Monitoring, measurement, analysis and evaluation

### 9.1.1 General

The organization shall monitor, measure, analyse and evaluate its environmental performance.

d) 実行可能な場合には，計画した対応処置を定期的にテストする．

e) 定期的に，また特に緊急事態の発生後又はテストの後には，プロセス及び計画した対応処置をレビューし，改訂する．

f) 必要に応じて，緊急事態への準備及び対応についての関連する情報及び教育訓練を，組織の管理下で働く人々を含む関連する利害関係者に提供する．

　組織は，プロセスが計画どおりに実施されるという確信をもつために必要な程度の，文書化した情報を維持しなければならない．

# 9　パフォーマンス評価
## 9.1　監視，測定，分析及び評価

### 9.1.1　一般
　組織は，環境パフォーマンスを監視し，測定し，分析し，評価しなければならない．

The organization shall determine:

a) what needs to be monitored and measured;

b) the methods for monitoring, measurement, analysis and evaluation, as applicable, to ensure valid results;

c) the criteria against which the organization will evaluate its environmental performance, and appropriate indicators;

d) when the monitoring and measuring shall be performed;

e) when the results from monitoring and measurement shall be analysed and evaluated.

The organization shall ensure that calibrated or verified monitoring and measurement equipment is used and maintained, as appropriate.

The organization shall evaluate its environmental performance and the effectiveness of the environmental management system.

The organization shall communicate relevant environmental performance information both internally and externally, as identified in its communica-

## 9 パフォーマンス評価　129

組織は，次の事項を決定しなければならない．

**a)** 監視及び測定が必要な対象

**b)** 該当する場合には，必ず，妥当な結果を確実に
するための，監視，測定，分析及び評価の方法

**c)** 組織が環境パフォーマンスを評価するための基
準及び適切な指標

**d)** 監視及び測定の実施時期

**e)** 監視及び測定の結果の，分析及び評価の時期


組織は，必要に応じて，校正された又は検証され
た監視機器及び測定機器が使用され，維持されてい
ることを確実にしなければならない．

組織は，環境パフォーマンス及び環境マネジメン
トシステムの有効性を評価しなければならない．


組織は，コミュニケーションプロセスで特定した
とおりに，かつ，順守義務による要求に従って，関
連する環境パフォーマンス情報について，内部と外

tion process(es) and as required by its compliance obligations.

The organization shall retain appropriate documented information as evidence of the monitoring, measurement, analysis and evaluation results.

### 9.1.2 Evaluation of compliance

The organization shall establish, implement and maintain the process(es) needed to evaluate fulfilment of its compliance obligations.

The organization shall:

a)  determine the frequency that compliance will be evaluated;

b)  evaluate compliance and take action if needed;

c)  maintain knowledge and understanding of its compliance status.

The organization shall retain documented information as evidence of the compliance evaluation result(s).

部の双方のコミュニケーションを行わなければならない.

　組織は，監視，測定，分析及び評価の結果の証拠として，適切な文書化した情報を保持しなければならない.

### 9.1.2　順守評価
　組織は，順守義務を満たしていることを評価するために必要なプロセスを確立し，実施し，維持しなければならない.

　組織は，次の事項を行わなければならない.
**a)**　順守を評価する頻度を決定する.

**b)**　順守を評価し，必要な場合には，処置をとる.

**c)**　順守状況に関する知識及び理解を維持する.


　組織は，順守評価の結果の証拠として，文書化した情報を保持しなければならない.

## 9.2 Internal audit

### 9.2.1 General

The organization shall conduct internal audits at planned intervals to provide information on whether the environmental management system:

a) conforms to:

    1) the organization's own requirements for its environmental management system;

    2) the requirements of this International Standard;

b) is effectively implemented and maintained.

### 9.2.2 Internal audit programme

The organization shall establish, implement and maintain (an) internal audit programme(s), including the frequency, methods, responsibilities, planning requirements and reporting of its internal audits.

When establishing the internal audit programme, the organization shall take into consideration the environmental importance of the processes concerned, changes affecting the organization and the

## 9.2 内部監査

### 9.2.1 一般

組織は，環境マネジメントシステムが次の状況にあるか否かに関する情報を提供するために，あらかじめ定めた間隔で内部監査を実施しなければならない．

**a)** 次の事項に適合している．

   **1)** 環境マネジメントシステムに関して，組織自体が規定した要求事項

   **2)** この規格の要求事項

**b)** 有効に実施され，維持されている．

### 9.2.2 内部監査プログラム

組織は，内部監査の頻度，方法，責任，計画要求事項及び報告を含む，内部監査プログラムを確立し，実施し，維持しなければならない．

内部監査プログラムを確立するとき，組織は，関連するプロセスの環境上の重要性，組織に影響を及ぼす変更及び前回までの監査の結果を考慮に入れなければならない．

134 ISO 14001

results of previous audits.

The organization shall:

a) define the audit criteria and scope for each audit;

b) select auditors and conduct audits to ensure objectivity and the impartiality of the audit process;

c) ensure that the results of the audits are reported to relevant management.

The organization shall retain documented information as evidence of the implementation of the audit programme and the audit results.

## 9.3 Management review

Top management shall review the organization's environmental management system, at planned intervals, to ensure its continuing suitability, adequacy and effectiveness.

The management review shall include consideration of:

組織は，次の事項を行わなければならない．

**a)** 各監査について，監査基準及び監査範囲を明確にする．

**b)** 監査プロセスの客観性及び公平性を確保するために，監査員を選定し，監査を実施する．

**c)** 監査の結果を関連する管理層に報告することを確実にする．

組織は，監査プログラムの実施及び監査結果の証拠として，文書化した情報を保持しなければならない．

## 9.3 マネジメントレビュー

トップマネジメントは，組織の環境マネジメントシステムが，引き続き，適切，妥当かつ有効であることを確実にするために，あらかじめ定めた間隔で，環境マネジメントシステムをレビューしなければならない．

マネジメントレビューは，次の事項を考慮しなければならない．

a) the status of actions from previous management reviews;

b) changes in:

    1) external and internal issues that are relevant to the environmental management system;

    2) the needs and expectations of interested parties, including compliance obligations;

    3) its significant environmental aspects;

    4) risks and opportunities;

c) the extent to which environmental objectives have been achieved;

d) information on the organization's environmental performance, including trends in:

    1) nonconformities and corrective actions;

    2) monitoring and measurement results;

    3) fulfilment of its compliance obligations;

    4) audit results;

e) adequacy of resources;

f) relevant communication(s) from interested parties, including complaints;

g) opportunities for continual improvement.

The outputs of the management review shall in-

9 パフォーマンス評価 137

**a)** 前回までのマネジメントレビューの結果とった
処置の状況

**b)** 次の事項の変化

  **1)** 環境マネジメントシステムに関連する外部及
び内部の課題

  **2)** 順守義務を含む，利害関係者のニーズ及び期
待

  **3)** 著しい環境側面

  **4)** リスク及び機会

**c)** 環境目標が達成された程度

**d)** 次に示す傾向を含めた，組織の環境パフォーマ
ンスに関する情報

  **1)** 不適合及び是正処置

  **2)** 監視及び測定の結果

  **3)** 順守義務を満たすこと

  **4)** 監査結果

**e)** 資源の妥当性

**f)** 苦情を含む，利害関係者からの関連するコミュ
ニケーション

**g)** 継続的改善の機会

マネジメントレビューからのアウトプットには，

138                     ISO 14001

clude:

— conclusions on the continuing suitability, adequacy and effectiveness of the environmental management system;

— decisions related to continual improvement opportunities;

— decisions related to any need for changes to the environmental management system, including resources;

— actions, if needed, when environmental objectives have not been achieved;

— opportunities to improve integration of the environmental management system with other business processes, if needed;

— any implications for the strategic direction of the organization.

The organization shall retain documented information as evidence of the results of management reviews.

## 10   Improvement
### 10.1   General
The organization shall determine opportunities for

次の事項を含めなければならない.

— 環境マネジメントシステムが,引き続き,適切,妥当かつ有効であることに関する結論

— 継続的改善の機会に関する決定

— 資源を含む,環境マネジメントシステムの変更の必要性に関する決定

— 必要な場合には,環境目標が達成されていない場合の処置
— 必要な場合には,他の事業プロセスへの環境マネジメントシステムの統合を改善するための機会
— 組織の戦略的な方向性に関する示唆

組織は,マネジメントレビューの結果の証拠として,文書化した情報を保持しなければならない.

## 10 改善
### 10.1 一般
組織は,環境マネジメントシステムの意図した成

improvement (see 9.1, 9.2 and 9.3) and implement necessary actions to achieve the intended outcomes of its environmental management system.

## 10.2 Nonconformity and corrective action

When a nonconformity occurs, the organization shall:

a)  react to the nonconformity and, as applicable:

   1)  take action to control and correct it;

   2)  deal with the consequences, including mitigating adverse environmental impacts;

b)  evaluate the need for action to eliminate the causes of the nonconformity, in order that it does not recur or occur elsewhere, by:

   1)  reviewing the nonconformity;

   2)  determining the causes of the nonconformity;

   3)  determining if similar nonconformities exist, or could potentially occur;

c)  implement any action needed;

果を達成するために，改善の機会（**9.1**，**9.2** 及び
**9.3** 参照）を決定し，必要な取組みを実施しなければならない．

## 10.2　不適合及び是正処置

不適合が発生した場合，組織は，次の事項を行わなければならない．

**a)** その不適合に対処し，該当する場合には，必ず，次の事項を行う．

  **1)** その不適合を管理し，修正するための処置をとる．

  **2)** 有害な環境影響の緩和を含め，その不適合によって起こった結果に対処する．

**b)** その不適合が再発又は他のところで発生しないようにするため，次の事項によって，その不適合の原因を除去するための処置をとる必要性を評価する．

  **1)** その不適合をレビューする．

  **2)** その不適合の原因を明確にする．

  **3)** 類似の不適合の有無，又はそれが発生する可能性を明確にする．

**c)** 必要な処置を実施する．

d) review the effectiveness of any corrective action taken;

e) make changes to the environmental management system, if necessary.

Corrective actions shall be appropriate to the significance of the effects of the nonconformities encountered, including the environmental impact(s).

The organization shall retain documented information as evidence of:

— the nature of the nonconformities and any subsequent actions taken;

— the results of any corrective action.

## 10.3 Continual improvement

The organization shall continually improve the suitability, adequacy and effectiveness of the environmental management system to enhance environmental performance.

d) とった是正処置の有効性をレビューする.

e) 必要な場合には,環境マネジメントシステムの変更を行う.

是正処置は,環境影響も含め,検出された不適合のもつ影響の著しさに応じたものでなければならない.

組織は,次に示す事項の証拠として,文書化した情報を保持しなければならない.
— 不適合の性質及びそれに対してとった処置

— 是正処置の結果

## 10.3 継続的改善

組織は,環境パフォーマンスを向上させるために,環境マネジメントシステムの適切性,妥当性及び有効性を継続的に改善しなければならない.

## Annex A
(informative)

## Guidance on the use
## of this International Standard

### A.1 General

The explanatory information given in this annex is intended to prevent misinterpretation of the requirements contained in this International Standard. While this information addresses and is consistent with these requirements, it is not intended to add to, subtract from, or in any way modify them.

The requirements in this International Standard need to be viewed from a systems or holistic perspective. The user should not read a particular sentence or clause of this International Standard in isolation from other clauses. There is an interrelationship between the requirements in some clauses and the requirements in other clauses. For example, the organization needs to understand the relationship between the commitments in its en-

# 附属書 A
## （参考）
## この規格の利用の手引

### A.1　一般

　この附属書に記載する説明は，この規格に規定する要求事項の誤った解釈を防ぐことを意図している．この情報は，この規格の要求事項と対応し整合しているが，要求事項に対して追加，削除，又は何らの変更を行うことも意図していない．

　この規格の要求事項は，システム又は包括的な観点から見る必要がある．利用者は，この規格の特定の文又は箇条を他の箇条と切り離して読まないほうがよい．箇条によっては，その箇条の要求事項と他の箇条の要求事項との間に相互関係があるものもある．例えば，組織は，環境方針におけるコミットメントと他の箇条で規定された要求事項との関係を理解する必要がある．

vironmental policy and the requirements that are specified in other clauses.

Management of change is an important part of maintaining the environmental management system that ensures the organization can achieve the intended outcomes of its environmental management system on an ongoing basis. Management of change is addressed in various requirements of this International Standard, including

— maintaining the environmental management system (see 4.4),

— environmental aspects (see 6.1.2),

— internal communication (see 7.4.2),

— operational control (see 8.1),

— internal audit programme (see 9.2.2), and

— management review (see 9.3).

As part of managing change, the organization should address planned and unplanned changes to ensure that the unintended consequences of these changes do not have a negative effect on the intended outcomes of the environmental management system. Examples of change include:

変更のマネジメントは，組織が継続して環境マネジメントシステムの意図した成果を達成できることを確実にする，環境マネジメントシステムの維持の重要な部分である．変更のマネジメントは，次を含むこの規格の様々な要求事項において規定されている．

— 環境マネジメントシステムの維持（**4.4** 参照）

— 環境側面（**6.1.2** 参照）
— 内部コミュニケーション（**7.4.2** 参照）
— 運用管理（**8.1** 参照）
— 内部監査プログラム（**9.2.2** 参照）
— マネジメントレビュー（**9.3** 参照）

変更のマネジメントの一環として，組織は，計画した変更及び計画していない変更について，それらの変更による意図しない結果が環境マネジメントシステムの意図した成果に好ましくない影響を与えないことを確実にするために，取り組むことが望ましい．変更の例には，次の事項が含まれる．

148 ISO 14001

— planned changes to products, processes, operations, equipment or facilities;
— changes in staff or external providers, including contractors;
— new information related to environmental aspects, environmental impacts and related technologies;
— changes in compliance obligations.

## A.2 Clarification of structure and terminology

The clause structure and some of the terminology of this International Standard have been changed to improve alignment with other management systems standards. There is, however, no requirement in this International Standard for its clause structure or terminology to be applied to an organization's environmental management system documentation. There is no requirement to replace the terms used by an organization with the terms used in this International Standard. Organizations can choose to use terms that suit their business, e.g. "records", "documentation", or "protocols", rather than "documented information".

附属書 A（参考） 149

— 製品，プロセス，運用，設備又は施設への，計画した変更

— スタッフの変更，又は請負者を含む外部提供者の変更

— 環境側面，環境影響及び関連する技術に関する新しい情報

— 順守義務の変化

## A.2 構造及び用語の明確化

この規格の箇条の構造及び一部の用語は，他のマネジメントシステム規格との一致性を向上させるために，旧規格から変更している．しかし，この規格では，組織の環境マネジメントシステムの文書にこの規格の箇条の構造又は用語を適用することは要求していない．組織が用いる用語をこの規格で用いている用語に置き換えることも要求していない．組織は，"文書化した情報"ではなく，"記録"，"文書類"又は"プロトコル"を用いるなど，それぞれの事業に適した用語を用いることを選択できる．

## A.3 Clarification of concepts

In addition to the terms and definitions given in Clause 3, clarification of selected concepts is provided below to prevent misunderstanding.

— In this International Standard, the use of the word "any" implies selection or choice.

— The words "appropriate" and "applicable" are not interchangeable. "Appropriate" means suitable (for, to) and implies some degree of freedom, while "applicable" means relevant or possible to apply and implies that if it can be done, it needs to be done.

— The word "consider" means it is necessary to think about the topic but it can be excluded; whereas "take into account" means it is necessary to think about the topic but it cannot be

附属書 A（参考）　　　151

## A.3　概念の明確化

箇条 3 に規定した用語及び定義のほかに，誤った解釈を防ぐために，幾つかの概念の説明を次に示す．

― この規格では，英語の "any" という言葉を用いる場合には，選定又は選択を意味している．

　　　**注記**　**JIS** では，英語の "any" は，"どのような" 又は "様々な" と訳しているほか，訳出していない場合もある．

― "適切な"，"必要に応じて" など（appropriate）と，"適用される"，"適用できる"，"該当する場合には，必ず" など（applicable）との間には，互換性はない．前者は，適している（suitable）という意味をもち，一定の自由度がある．後者は，関連する，又は適用することが可能である，という意味をもち，可能な場合には行う必要がある，という意味を含んでいる．

― "考慮する"（consider）という言葉は，その事項について考える必要があるが除外することができる，という意味をもつ．他方，"考慮に入れる"（take into account）は，その事項につ

excluded.

— "Continual" indicates duration that occurs over a period of time, but with intervals of interruption (unlike "continuous" which indicates duration without interruption). "Continual" is therefore the appropriate word to use when referring to improvement.
— In this International Standard, the word "effect" is used to describe the result of a change to the organization. The phrase "environmental impact" refers specifically to the result of a change to the environment.
— The word "ensure" means the responsibility can be delegated, but not the accountability.

— This International Standard uses the term "interested party"; the term "stakeholder" is a synonym as it represents the same concept.

This International Standard uses some new terminology. A brief explanation is given below to aid

いて考える必要があり，かつ，除外できない，という意味をもつ．

— "継続的"（continual）とは，一定の期間にわたって続くことを意味しているが，途中に中断が入る［中断なく続くことを意味する"連続的"（continuous）とは異なる．］．したがって，改善について言及する場合には，"継続的"という言葉を用いるのが適切である．

— この規格では，"影響"（effect）という言葉は，組織に対する変化の結果を表すために用いている．"環境影響"（environmental impact）という表現は，特に，環境に対する変化の結果を意味している．

— "確実にする"及び"確保する"（ensure）という言葉は，責任を委譲することができるが，説明責任については委譲できないことを意味する．

— この規格では，"利害関係者"（interested party）という用語を用いている．"ステークホルダー"（stakeholder）という用語は，同じ概念を表す同義語である．

この規格では，幾つかの新しい用語を用いている．この規格の新規の利用者及び旧規格の利用者の

both new users and those who have used previous editions of this International Standard.

— The phrase "compliance obligations" replaces the phrase "legal requirements and other requirements to which the organization subscribes" used in the previous edition of this International Standard. The intent of this new phrase does not differ from that of the previous edition.

— "Documented information" replaces the nouns "documentation", "documents" and "records" used in previous editions of this International Standard. To distinguish the intent of the generic term "documented information", this International Standard now uses the phrase "retain documented information as evidence of...." to mean records, and "maintain documented information" to mean documentation other than records. The phrase "as evidence of...." is not a requirement to meet legal evidentiary requirements; its intent is only to indicate objective evidence needs to be retained.

— The phrase "external provider" means an external supplier organization (including a con-

附属書 A（参考）　　　155

双方の助けとなるよう，これらの用語についての簡単な説明を次に示す．

— "順守義務" という表現は，旧規格で用いていた "法的要求事項及び組織が同意するその他の要求事項" という表現に置き換わるものである．この新しい表現の意味は，旧規格から変更していない．

— "文書化した情報" は，旧規格で用いていた "文書類"，"文書" 及び "記録" という名詞に置き換わるものである．一般用語としての "文書化した情報" の意図と区別するため，この規格では，記録を意味する場合には "…の証拠として，文書化した情報を保持する" という表現を用い，記録以外の文書類を意味する場合には "文書化した情報を維持する" という表現を用いている．"…の証拠として" という表現は，法的な証拠となる要求事項を満たすことの要求ではなく，保持する必要がある客観的証拠を示すことだけを意図している．

— "外部提供者" という表現は，製品又はサービスを提供する外部供給者の組織（請負者を含

tractor) that provides a product or a service.

— The change from "identify" to "determine" is intended to harmonize with the standardized management system terminology. The word "determine" implies a discovery process that results in knowledge. The intent does not differ from that of previous editions.

— The phrase "intended outcome" is what the organization intends to achieve by implementing its environmental management system. The minimal intended outcomes include enhancement of environmental performance, fulfilment of compliance obligations and achievement of environmental objectives. Organizations can set additional intended outcomes for their environmental management system. For example, consistent with their commitment to protection of the environment, an organization may establish an intended outcome to work towards sustainable development.

— The phrase "person(s) doing work under its control" includes persons working for the or-

む.）を意味する.

— "特定する"（identify）から，"決定する"など（determine）に変更した意図は，標準化されたマネジメントシステムの用語と一致させるためである．"決定する"など（determine）という言葉は，知識をもたらす発見のプロセスを意味している．その意味は，旧規格から変更していない．

— "意図した成果"（intended outcome）という表現は，組織が環境マネジメントシステムの実施によって達成しようとするものである．最低限の意図した成果には，環境パフォーマンスの向上，順守義務を満たすこと，及び環境目標の達成が含まれる．組織は，それぞれの環境マネジメントシステムについて，追加の意図した成果を設定することができる．例えば，環境保護へのコミットメントと整合して，組織は，持続可能な開発に取り組むための意図した成果を確立してもよい．

— "組織の管理下で働く人（又は人々）"という表現は，組織で働く人々，及び組織が責任をも

ganization and those working on its behalf for which the organization has responsibility (e.g. contractors). It replaces the phrase "persons working for it or on its behalf" and "persons working for or on behalf of the organization" used in the previous edition of this International Standard. The intent of this new phrase does not differ from that of the previous edition.

— The concept of "target" used in previous editions of this International Standard is captured within the term "environmental objective".

## A.4 Context of the organization

## A.4.1 Understanding the organization and its context

The intent of 4.1 is to provide a high-level, conceptual understanding of the important issues that can affect, either positively or negatively, the way the organization manages its environmental responsibilities. Issues are important topics for the organization, problems for debate and discussion or changing circumstances that affect the organi-

つ，組織のために働く人々（例えば，請負者）を含む．この表現は，旧規格で用いていた"組織で働く又は組織のために働く人"という表現に置き換わるものである．この新しい表現の意味は，旧規格から変更していない．

― 旧規格で用いていた"目標"（target）の概念は，"環境目標"（environmental objective）の用語の中に包含されている．

## A.4 組織の状況
### A.4.1 組織及びその状況の理解

4.1 は，組織が自らの環境責任をマネジメントする方法に対して好ましい又は好ましくない影響を与える可能性のある重要な課題についての，高いレベルでの，概念的な理解を提供することを意図している．課題とは，組織にとって重要なトピック，討議及び議論のための問題，又は環境マネジメントシステムに関して設定した意図した成果を達成する組織

zation's ability to achieve the intended outcomes it sets for its environmental management system.

Examples of internal and external issues which can be relevant to the context of the organization include:

a)  environmental conditions related to climate, air quality, water quality, land use, existing contamination, natural resource availability and biodiversity, that can either affect the organization's purpose, or be affected by its environmental aspects;

b)  the external cultural, social, political, legal, regulatory, financial, technological, economic, natural and competitive circumstances, whether international, national, regional or local;

c)  the internal characteristics or conditions of the organization, such as its activities, products and services, strategic direction, culture and capabilities (i.e. people, knowledge, processes, systems).

An understanding of the context of an organiza-

の能力に影響を与える，変化している周囲の状況である．

組織の状況に関連し得る内部及び外部の課題の例には，次の事項を含む．

a) 気候，大気の質，水質，土地利用，既存の汚染，天然資源の利用可能性及び生物多様性に関連した環境状態で，組織の目的に影響を与える可能性のある，又は環境側面によって影響を受ける可能性のあるもの

b) 国際，国内，地方又は近隣地域を問わず，外部の文化，社会，政治，法律，規制，金融，技術，経済，自然及び競争の状況

c) 組織の活動，製品及びサービス，戦略的な方向性，文化，能力（すなわち，人々，知識，プロセス及びシステム）などの，組織の内部の特性又は状況

組織の状況の理解は，環境マネジメントシステム

tion is used to establish, implement, maintain and continually improve its environmental management system (see 4.4). The internal and external issues that are determined in 4.1 can result in risks and opportunities to the organization or to the environmental management system (see 6.1.1 to 6.1.3). The organization determines those that need to be addressed and managed (see 6.1.4, 6.2, Clause 7, Clause 8 and 9.1).

## A.4.2 Understanding the needs and expectations of interested parties

An organization is expected to gain a general (i.e. high-level, not detailed) understanding of the expressed needs and expectations of those internal and external interested parties that have been determined by the organization to be relevant. The organization considers the knowledge gained when determining which of these needs and expectations it has to or it chooses to comply with, i.e. its compliance obligations (see 6.1.1).

In the case of an interested party perceiving itself to be affected by the organization's decisions or ac-

附属書 A（参考）　　163

を確立し，実施し，維持し，継続的に改善するために用いられる（**4.4** 参照）．**4.1** で決定した内部及び外部の課題は，組織又は環境マネジメントシステムに対するリスク及び機会をもたらし得る（**6.1.1 〜 6.1.3** 参照）．組織は，取り組み，マネジメントする必要がある（**6.1.4**，**6.2**，箇条 **7**，箇条 **8** 及び **9.1** を参照．）リスク及び機会を決定する．

## A.4.2　利害関係者のニーズ及び期待の理解

　組織は，関連すると決定した内部及び外部の利害関係者から表明されたニーズ及び期待についての一般的な（すなわち，詳細ではなく，高いレベルで）理解を得ることが期待されている．組織は，得たその知識を，これらのニーズ及び期待の中から順守しなければならない又は順守することを選ぶもの，すなわち組織の順守義務となるものを決定するときに，考慮することとなる（**6.1.1** 参照）．

　利害関係者が，環境パフォーマンスに関連する組織の決定又は活動に影響を受けると認識している場

tivities related to environmental performance, the organization considers the relevant needs and expectations that are made known or have been disclosed by the interested party to the organization.

Interested party requirements are not necessarily requirements of the organization. Some interested party requirements reflect needs and expectations that are mandatory because they have been incorporated into laws, regulations, permits and licences by governmental or even court decision. The organization may decide to voluntarily agree to or adopt other requirements of interested parties (e.g. entering into a contractual relationship, subscribing to a voluntary initiative). Once the organization adopts them, they become organizational requirements (i.e. compliance obligations) and are taken into account when planning the environmental management system (see 4.4). A more detailed-level analysis of its compliance obligations is performed in 6.1.3.

## A.4.3 Determining the scope of the environmental management system

合には，組織は，その利害関係者によって組織に知らされている又は開示されている，関連するニーズ及び期待を考慮することとなる．

利害関係者の要求事項は，必ずしも組織の要求事項になるわけではない．利害関係者の要求事項の中には，政府又は裁判所の判決によって，法令，規制，許可及び認可の中に導入されていることで強制的になっているニーズ及び期待を反映しているものもある．組織は，利害関係者のその他の要求事項について，自発的に合意又は採用することを決めてもよい（例えば，契約関係の締結，自発的取組みの合意）．組織が採用したものは，組織の要求事項，すなわち，順守義務となり，環境マネジメントシステムを計画するときに考慮に入れることとなる（4.4参照）．より詳細なレベルでの順守義務の分析は，6.1.3 で実施される．

## A.4.3 環境マネジメントシステムの適用範囲の決定

The scope of the environmental management system is intended to clarify the physical and organizational boundaries to which the environmental management system applies, especially if the organization is a part of a larger organization. An organization has the freedom and flexibility to define its boundaries. It may choose to implement this International Standard throughout the entire organization, or only in (a) specific part(s) of the organization, as long as the top management for that (those) part(s) has authority to establish an environmental management system.

In setting the scope, the credibility of the environmental management system depends upon the choice of organizational boundaries. The organization considers the extent of control or influence that it can exert over activities, products and services considering a life cycle perspective. Scoping should not be used to exclude activities, products, services, or facilities that have or can have significant environmental aspects, or to evade its compliance obligations. The scope is a factual and representative statement of the organization's operations

附属書 A（参考）

　環境マネジメントシステムの適用範囲の意図は，環境マネジメントシステムが適用される物理的及び組織上の境界を明確にすることであり，特にその組織がより大きい組織の一部である場合にはそれが必要である．組織は，その境界を定める自由度及び柔軟性をもつ．組織は，この規格を組織全体に実施するか，又は組織の特定の一部（複数の場合もある．）だけにおいて，その部分のトップマネジメントが環境マネジメントシステムを確立する権限をもつ限りにおいて，その部分に対して実施するかを選択してもよい．


　適用範囲の設定において，環境マネジメントシステムへの信ぴょう（憑）性は，どのように組織上の境界を選択するかによって決まる．組織は，ライフサイクルの視点を考慮して，活動，製品及びサービスに対して管理できる又は影響を及ぼすことができる程度を検討することとなる．適用範囲の設定を，著しい環境側面をもつ若しくはもつ可能性のある活動・製品・サービス・施設を除外するため，又は順守義務を逃れるために用いないほうがよい．適用範囲は，事実に基づくもので，環境マネジメントシステムの境界内に含まれる組織の運用を表した記述で

included within its environmental management system boundaries that should not mislead interested parties.

Once the organization asserts it conforms to this International Standard, the requirement to make the scope statement available to interested parties applies.

## A.4.4 Environmental management system

The organization retains authority and accountability to decide how it fulfils the requirements of this International Standard, including the level of detail and extent to which it:

a) establishes one or more processes to have confidence that it (they) is (are) controlled, carried out as planned and achieve the desired results;

b) integrates environmental management system requirements into its various business processes, such as design and development, procurement, human resources, sales and marketing;

c) incorporates issues associated with the con-

あり，その記述は，利害関係者の誤解を招かないものであることが望ましい．

この規格への適合を宣言すると，適用範囲の記述を利害関係者に対して入手可能にすることの要求事項が適用される．

## A.4.4　環境マネジメントシステム

組織は，次の事項を実施するに当たっての詳細さのレベル及び程度を含む，この規格の要求事項を満たす方法を決定する権限及び説明責任を保持している．

a)　そのプロセスが管理され，計画どおりに実施され，望ましい結果を達成しているという確信をもつために，一つ又は複数のプロセスを確立する．

b)　設計及び開発，調達，人的資源，販売，マーケティングなどの種々の事業プロセスに，環境マネジメントシステム要求事項を統合する．

c)　組織の状況に関する課題（4.1 参照）及び利害

text of the organization (see 4.1) and interested party requirements (see 4.2) within its environmental management system.

If this International Standard is implemented for (a) specific part(s) of an organization, policies, processes and documented information developed by other parts of the organization can be used to meet the requirements of this International Standard, provided they are applicable to that (those) specific part(s).

For information on maintaining the environmental management system as part of management of change, see Clause A.1.

## A.5  Leadership
### A.5.1  Leadership and commitment
To demonstrate leadership and commitment, there are specific responsibilities related to the environmental management system in which top management should be personally involved or which top management should direct. Top management may delegate responsibility for these actions to others,

附属書A（参考）

関係者の要求事項（**4.2**参照）を，環境マネジメントシステムの中に組み込む．

組織の特定の一部（複数の場合もある．）に対してこの規格を実施する場合には，組織の他の部分が策定した方針，プロセス及び文書化した情報がその特定の一部にも適用可能であれば，この規格の要求事項を満たすものとしてそれらの方針，プロセス及び文書化した情報を用いることができる．

変更のマネジメントの一部としての環境マネジメントシステムの維持に関する情報を，**A.1**に示す．

## A.5　リーダーシップ

### A.5.1　リーダーシップ及びコミットメント

リーダーシップ及びコミットメントを実証するために，トップマネジメント自身が関与又は指揮することが望ましい，環境マネジメントシステムに関連する特定の責任がある．トップマネジメントは，他の人にこれらの行動の責任を委譲してもよいが，それらが実施されたことを確実にすることに対する説

but it retains accountability for ensuring the actions are performed.

## A.5.2 Environmental policy

An environmental policy is a set of principles stated as commitments in which top management outlines the intentions of the organization to support and enhance its environmental performance. The environmental policy enables the organization to set its environmental objectives (see 6.2), take actions to achieve the intended outcomes of the environmental management system, and achieve continual improvement (see Clause 10).

Three basic commitments for the environmental policy are specified in this International Standard to:

a)    protect the environment;

b)    fulfil the organization's compliance obligations;

c)    continually improve the environmental management system to enhance environmental performance.

明責任は，トップマネジメントが保持する．

## A.5.2 環境方針

環境方針は，環境パフォーマンスを支え，向上させるために，トップマネジメントが組織の意図を示すコミットメントとして明示する，一連の原則である．環境方針によって，組織は，環境目標を設定し（**6.2** 参照），環境マネジメントシステムの意図した成果を達成するために取り組み，継続的改善を達成する（箇条 **10** 参照）ことが可能となる．

この規格は，次に示す，環境方針の三つの基本的なコミットメントを規定している．

**a)** 環境を保護する．

**b)** 組織の順守義務を満たす．

**c)** 環境パフォーマンスを向上させるために，環境マネジメントシステムを継続的に改善する．

These commitments are then reflected in the processes an organization establishes to address specific requirements in this International Standard, to ensure a robust, credible and reliable environmental management system.

The commitment to protect the environment is intended to not only prevent adverse environmental impacts through prevention of pollution, but to protect the natural environment from harm and degradation arising from the organization's activities, products and services. The specific commitment(s) an organization pursues should be relevant to the context of the organization, including the local or regional environmental conditions. These commitments can address, for example, water quality, recycling, or air quality, and can also include commitments related to climate change mitigation and adaptation, protection of biodiversity and ecosystems, and restoration.

While all the commitments are important, some interested parties are especially concerned with the organization's commitment to fulfil its compli-

附属書A（参考）

これらのコミットメントは，しっかりとした，信ぴょう（憑）性及び信頼性のある環境マネジメントシステムを確実にするため，組織がこの規格の特定の要求事項に取り組むために確立するプロセスに反映されることとなる．

環境保護へのコミットメントは，汚染の予防を通じて有害な環境影響を防止することだけでなく，組織の活動，製品及びサービスから生じる危害及び劣化から自然環境を保護することも意図している．組織が追求する固有のコミットメントは，近隣地域又は地方の環境状態を含む，組織の状況に関連するものであることが望ましい．これらのコミットメントは，水質，リサイクル，大気の質などに取り組むものであることもあれば，気候変動の緩和及び気候変動への適応，生物多様性及び生態系の保護，並びに回復に関連したコミットメントを含むこともある．

全てのコミットメントが重要ではあるが，利害関係者には，順守義務，中でも適用される法的要求事項を満たすことに対する組織のコミットメントに，

ance obligations, particularly applicable legal requirements. This International Standard specifies a number of interconnected requirements related to this commitment. These include the need to:

— determine compliance obligations;
— ensure operations are carried out in accordance with these compliance obligations;
— evaluate fulfilment of the compliance obligations;
— correct nonconformities.

## A.5.3 Organizational roles, responsibilities and authorities

Those involved in the organization's environmental management system should have a clear understanding of their role, responsibility(ies) and authority(ies) for conforming to the requirements of this International Standard and achieving the intended outcomes.

The specific roles and responsibilities identified in 5.3 may be assigned to an individual, sometimes referred to as the "management representative", shared by several individuals, or assigned to a

附属書 A（参考）　　　177

特に関心をもつ者もいる．この規格では，このコミットメントに関連した，多くの相互に関連する要求事項を規定している．これらには，次の事項についての必要性が含まれる．

— 順守義務を決定する．
— それらの順守義務に従って運用が行われていることを確実にする．
— 順守義務を満たしていることを評価する．

— 不適合を修正する．

## A.5.3　組織の役割，責任及び権限

組織の環境マネジメントシステムに関与する人々は，この規格の要求事項への適合及び意図した成果の達成に関する，自らの役割，責任及び権限について，明確に理解していることが望ましい．

**5.3** で特定した役割及び責任は，"管理責任者"と呼ばれることもある個人に割り当てても，複数の人々で分担しても，又はトップマネジメントのメンバーに割り当ててもよい．

member of top management.

## A.6    Planning

### A.6.1    Actions to address risks and opportunities

#### A.6.1.1    General

The overall intent of the process(es) established in 6.1.1 is to ensure that the organization is able to achieve the intended outcomes of its environmental management system, to prevent or reduce undesired effects, and to achieve continual improvement. The organization can ensure this by determining its risks and opportunities that need to be addressed and planning action to address them. These risks and opportunities can be related to environmental aspects, compliance obligations, other issues or other needs and expectations of interested parties.

Environmental aspects (see 6.1.2) can create risks and opportunities associated with adverse environmental impacts, beneficial environmental impacts, and other effects on the organization. The risks and opportunities related to environmental

## A.6　計画
## A.6.1　リスク及び機会への取組み

### A.6.1.1　一般

**6.1.1** で確立されるプロセスの全体的な意図は，組織が環境マネジメントシステムの意図した成果を達成し，望ましくない影響を防止又は低減し，継続的改善を達成できることを確実にすることである．組織は，これらのことを，取り組む必要があるリスク及び機会を決定し，それらへの取組みを計画することによって確実にすることができる．これらのリスク及び機会は，環境側面，順守義務，その他の課題，又は利害関係者のその他のニーズ及び期待に関連し得る．

環境側面（**6.1.2** 参照）は，有害な環境影響，有益な環境影響，及び組織に対するその他の影響に関連する，リスク及び機会を生み出し得る．環境側面に関連するリスク及び機会は，著しさの評価の一部として決定することも，又は個別に決定することも

180                    ISO 14001

aspects can be determined as part of the significance evaluation or determined separately.

Compliance obligations (see 6.1.3) can create risks and opportunities, such as failing to comply (which can damage the organization's reputation or result in legal action) or performing beyond its compliance obligations (which can enhance the organization's reputation).

The organization can also have risks and opportunities related to other issues, including environmental conditions or needs and expectations of interested parties, which can affect the organization's ability to achieve the intended outcomes of its environmental management system, e.g.

a)  environmental spillage due to literacy or language barriers among workers who cannot understand local work procedures;

b)  increased flooding due to climate change that could affect the organizations premises;

c)  lack of available resources to maintain an effective environmental management system due to economic constraints;

できる.

順守義務（**6.1.3** 参照）は，不順守（これは，組織の評判を害し得る，又は法的行動につながり得る.），順守義務を超えた実施（これは，組織の評判の強化につながり得る.）のような，リスク及び機会を生み出し得る.

組織は，また，環境マネジメントシステムの意図した成果を達成する組織の能力に影響を与え得る，環境状態又は利害関係者のニーズ及び期待を含むその他の課題に関連する，リスク及び機会をもち得る．こうしたリスク及び機会の例には，次に示すものがある.

**a)** 労働者間の識字又は言葉の壁によって現地の業務手順を理解できないことによる，環境への流出.

**b)** 組織の構内に影響を与え得る，気候変動による洪水の増加

**c)** 経済的制約による，有効な環境マネジメントシステムを維持するための利用可能な資源の欠如

d) introducing new technology financed by governmental grants, which could improve air quality;

e) water scarcity during periods of drought that could affect the organization's ability to operate its emission control equipment.

Emergency situations are unplanned or unexpected events that need the urgent application of specific competencies, resources or processes to prevent or mitigate their actual or potential consequences. Emergency situations can result in adverse environmental impacts or other effects on the organization. When determining potential emergency situations (e.g. fire, chemical spill, severe weather), the organization should consider:

— the nature of onsite hazards (e.g. flammable liquids, storage tanks, compressed gasses);

— the most likely type and scale of an emergency situation;

— the potential for emergency situations at a nearby facility (e.g. plant, road, railway line).

Although risks and opportunities need to be de-

**d)** 大気の質を改善し得る，政府の助成を利用した新しい技術の導入

**e)** 排出管理設備を運用する組織の能力に影響を与え得る，干ばつ期における水不足

　緊急事態は，顕在した又は潜在的な結果を防止又は緩和するために特定の力量，資源又はプロセスの緊急の適用を必要とする，計画していない又は予期しない事象である．緊急事態は，有害な環境影響又は組織に対するその他の影響をもたらす可能性がある．潜在的な緊急事態（例えば，火災，化学物質の漏えい，悪天候）を決定するとき，組織は，次の事項を考慮することが望ましい．

— 現場ハザードの性質（例えば，可燃性液体，貯蔵タンク，圧縮ガス）
— 緊急事態の最も起こりやすい種類及び規模

— 近接した施設（例えば，プラント，道路，鉄道）で緊急事態が発生する可能性

　リスク及び機会は，決定し，取り組む必要がある

termined and addressed, there is no requirement for formal risk management or a documented risk management process. It is up to the organization to select the method it will use to determine its risks and opportunities. The method may involve a simple qualitative process or a full quantitative assessment depending on the context in which the organization operates.

The risks and opportunities identified (see 6.1.1 to 6.1.3) are inputs for planning actions (see 6.1.4) and for establishing the environmental objectives (see 6.2).

### A.6.1.2   Environmental aspects

An organization determines its environmental aspects and associated environmental impacts, and determines those that are significant and, therefore, need to be addressed by its environmental management system.

Changes to the environment, either adverse or beneficial, that result wholly or partially from environmental aspects are called environmental

が，正式なリスクマネジメント又は文書化したリスクマネジメントプロセスは要求していない．リスク及び機会を決定するために用いる方法の選定は，組織に委ねられている．この方法には，組織の活動が行われる状況に応じて，単純な定性的プロセス又は完全な定量的評価を含めてもよい．

　特定されたリスク及び機会（**6.1.1～6.1.3** 参照）は，取組みの計画策定（**6.1.4** 参照）及び環境目標の確立（**6.2** 参照）へのインプットとなる．

## A.6.1.2　環境側面

　組織は，環境側面及びそれに伴う環境影響を決定し，それらのうち，環境マネジメントシステムによって取り組む必要がある著しいものを決定する．

　有害か有益かを問わず，全体的に又は部分的に環境側面から生じる，環境に対する変化を環境影響という．環境影響は，近隣地域，地方及び地球規模で

impacts. The environmental impact can occur at local, regional and global scales, and also can be direct, indirect or cumulative by nature. The relationship between environmental aspects and environmental impacts is one of cause and effect.

When determining environmental aspects, the organization considers a life cycle perspective. This does not require a detailed life cycle assessment; thinking carefully about the life cycle stages that can be controlled or influenced by the organization is sufficient. Typical stages of a product (or service) life cycle include raw material acquisition, design, production, transportation/delivery, use, end-of-life treatment and final disposal. The life cycle stages that are applicable will vary depending on the activity, product or service.

An organization needs to determine the environmental aspects within the scope of its environmental management system. It takes into account the inputs and outputs (both intended and unintended) that are associated with its current and relevant past activities, products and services;

附属書A（参考）

起こり得るものであり，また，直接的なもの，間接的なもの，又は性質上累積的なものでもあり得る．環境側面と環境影響との関係は，一種の因果関係である．

環境側面を決定するとき，組織は，ライフサイクルの視点を考慮する．これは，詳細なライフサイクルアセスメントを要求するものではなく，組織が管理できる又は影響を及ぼすことができるライフサイクルの段階について注意深く考えることで十分である．製品（又はサービス）の典型的なライフサイクルの段階には，原材料の取得，設計，生産，輸送又は配送（提供），使用，使用後の処理及び最終処分が含まれる．適用できるライフサイクルの段階は，活動，製品又はサービスによって異なる．

組織は，環境マネジメントシステムの適用範囲内にある環境側面を決定する必要がある．組織は，現在及び関連する過去の活動，製品及びサービス，計画した又は新規の開発，並びに新規の又は変更された活動，製品及びサービスに関係するインプット及びアウトプット（意図するか意図しないかにかかわ

planned or new developments; and new or modified activities, products and services. The method used should consider normal and abnormal operating conditions, shut-down and start-up conditions, as well as the reasonably foreseeable emergency situations identified in 6.1.1. Attention should be paid to prior occurrences of emergency situations. For information on environmental aspects as part of managing change, see Clause A.1.

An organization does not have to consider each product, component or raw material individually to determine and evaluate their environmental aspects; it may group or categorize activities, products and services when they have common characteristics.

When determining its environmental aspects, the organization can consider:

a)    emissions to air;

b)    releases to water;

c)    releases to land;

d)    use of raw materials and natural resources;

e)    use of energy;

附属書 A（参考）

らず）を考慮に入れる．用いる方法は，通常及び非通常の運用状況，停止及び立ち上げの状況，並びに **6.1.1** で特定した合理的に予見できる緊急事態を考慮することが望ましい．過去の緊急事態の発生について，注意を払うことが望ましい．変更のマネジメントの一部としての環境側面に関する情報を，**A.1** に示す．

組織は，環境側面を決定し評価するために，製品，部品又は原材料をそれぞれ個別に考慮する必要はなく，活動，製品及びサービスに共通の特性がある場合には，その活動，製品及びサービスをグループ化又は分類してもよい．

環境側面を決定するとき，組織は，次の事項を考慮することができる．
a) 大気への排出
b) 水への排出
c) 土地への排出
d) 原材料及び天然資源の使用
e) エネルギーの使用

190                        ISO 14001

f)   energy emitted (e.g. heat, radiation, vibration (noise), light);

g)   generation of waste and/or by-products;

h)   use of space.


In addition to the environmental aspects that it can control directly, an organization determines whether there are environmental aspects that it can influence. These can be related to products and services used by the organization which are provided by others, as well as products and services that it provides to others, including those associated with (an) outsourced process(es). With respect to those an organization provides to others, it can have limited influence on the use and end-of-life treatment of the products and services. In all circumstances, however, it is the organization that determines the extent of control it is able to exercise, the environmental aspects it can influence, and the extent to which it chooses to exercise such influence.


Consideration should be given to environmental aspects related to the organization's activities,

附属書 A（参考）　　　191

**f)** 排出エネルギー［例えば，熱，放射，振動（騒音），光］

**g)** 廃棄物及び／又は副産物の発生

**h)** 空間の使用

　組織は，組織が直接的に管理できる環境側面のほかに，影響を及ぼすことができる環境側面があるか否かを決定する．これは，他者から提供され，組織が使用する製品及びサービス，並びに組織が他者に提供する製品及びサービス（外部委託したプロセスに関連するものも含む．）に関連し得る．組織が他者に提供する製品及びサービスについて，組織は，その製品及びサービスの使用及び使用後の処理に対して限定された影響しかもつことができない場合がある．しかし，いかなる場合においても，組織が管理できる程度，影響を及ぼすことができる環境側面，及び組織が行使することを選択するそうした影響の程度を決定するのは，組織である．

　組織の活動，製品及びサービスに関係する環境側面の例として，次の事項を考慮することが望まし

products and services, such as:

— design and development of its facilities, processes, products and services;

— acquisition of raw materials, including extraction;

— operational or manufacturing processes, including warehousing;

— operation and maintenance of facilities, organizational assets and infrastructure;

— environmental performance and practices of external providers;

— product transportation and service delivery, including packaging;

— storage, use and end-of-life treatment of products;

— waste management, including reuse, refurbishing, recycling and disposal.

There is no single method for determining significant environmental aspects, however, the method and criteria used should provide consistent results. The organization sets the criteria for determining its significant environmental aspects. Environmental criteria are the primary and minimum

い.

— 施設，プロセス，製品及びサービスの設計及び
　　開発

— 採取を含む，原材料の取得

— 倉庫保管を含む，運用又は製造のプロセス

— 施設，組織の資産及びインフラストラクチャ
　　の，運用及びメンテナンス

— 外部提供者の環境パフォーマンス及び業務慣行

— 包装を含む，製品の輸送及びサービスの提供

— 製品の保管，使用及び使用後の処理

— 廃棄物管理．これには，再利用，修復，リサイ
　　クル及び処分を含む．

　著しい環境側面を決定する方法は，一つだけでは
ない．しかし，用いる方法及び基準は，矛盾のない
一貫した結果を出すものであることが望ましい．組
織は，著しい環境側面を決定するための基準を設定
する．環境に関する基準は，環境側面を評価するた
めの主要かつ最低限の基準である．基準は，環境側

criteria for assessing environmental aspects. Criteria can relate to the environmental aspect (e.g. type, size, frequency) or the environmental impact (e.g. scale, severity, duration, exposure). Other criteria may also be used. An environmental aspect might not be significant when only considering environmental criteria. It can, however, reach or exceed the threshold for determining significance when other criteria are considered. These other criteria can include organizational issues, such as legal requirements or interested party concerns. These other criteria are not intended to be used to downgrade an aspect that is significant based on its environmental impact.

A significant environmental aspect can result in one or more significant environmental impacts, and can therefore result in risks and opportunities that need to be addressed to ensure the organization can achieve the intended outcomes of its environmental management system.

## A.6.1.3 Compliance obligations

The organization determines, at a sufficiently de-

面（例えば，種類，規模，頻度）に関連することもあれば，環境影響（例えば，規模，深刻度，継続時間，暴露）に関連することもある．組織は，その他の基準を用いてもよい．ある環境側面は，環境に関する基準を考慮するだけの場合には著しくなかったとしても，その他の基準を考慮した場合には，著しさを決定するためのしきい（閾）値に達するか，又はそれを超える可能性がある．これらのその他の基準には，法的要求事項，利害関係者の関心事などの，組織の課題を含み得る．これらのその他の基準は，環境影響に基づいて著しさがある側面を過小評価するために用いられることを意図したものではない．

著しい環境側面は，一つ又は複数の著しい環境影響をもたらす可能性があるため，組織が環境マネジメントシステムの意図した成果を達成することを確実にするために取り組む必要があるリスク及び機会をもたらし得る．

## A.6.1.3 順守義務

組織は，**4.2** で特定した順守義務のうち環境側面

tailed level, the compliance obligations it identified in 4.2 that are applicable to its environmental aspects, and how they apply to the organization. Compliance obligations include legal requirements that an organization has to comply with and other requirements that the organization has to or chooses to comply with.

Mandatory legal requirements related to an organization's environmental aspects can include, if applicable:

a) requirements from governmental entities or other relevant authorities;

b) international, national and local laws and regulations;

c) requirements specified in permits, licenses or other forms of authorization;

d) orders, rules or guidance from regulatory agencies;

e) judgements of courts or administrative tribunals.

Compliance obligations also include other interested party requirements related to its environmen-

附属書 A（参考）

に適用されるもの，及びどのようにそれらの順守義務を組織に適用するかについての，十分に詳細なレベルでの決定を行う．順守義務には，組織が順守しなければならない法的要求事項，及び組織が順守しなければならない又は順守することを選んだその他の要求事項が含まれる．

　組織の環境側面に関連する強制的な法的要求事項には，適用可能な場合には，次が含まれ得る．

a)　政府機関又はその他の関連当局からの要求事項

b)　国際的な，国の及び近隣地域の法令及び規制

c)　許可，認可又はその他の承認の形式において規定される要求事項

d)　規制当局による命令，規則又は指針

e)　裁判所又は行政審判所の判決

　順守義務は，組織が採用しなければならない又は採用することを選ぶ，組織の環境マネジメントシス

tal management system which the organization has to or chooses to adopt. These can include, if applicable:

— agreements with community groups or non-governmental organizations;

— agreements with public authorities or customers;

— organizational requirements;

— voluntary principles or codes of practice;

— voluntary labelling or environmental commitments;

— obligations arising under contractual arrangements with the organization;

— relevant organizational or industry standards.

## A.6.1.4 Planning action

The organization plans, at a high level, the actions that have to be taken within the environmental management system to address its significant environmental aspects, its compliance obligations, and the risks and opportunities identified in 6.1.1 that are a priority for the organization to achieve the intended outcomes of its environmental management system.

テムに関連した，利害関係者のその他の要求事項も含む．これらには，適用可能な場合には，次が含まれ得る．

— コミュニティグループ又は非政府組織（NGO）との合意
— 公的機関又は顧客との合意

— 組織の要求事項
— 自発的な原則又は行動規範
— 自発的なラベル又は環境コミットメント

— 組織との契約上の取決めによって生じる義務

— 関連する，組織又は業界の標準

## A.6.1.4　取組みの計画策定

　組織は，組織が環境マネジメントシステムの意図した成果を達成するための優先事項である，著しい環境側面，順守義務，並びに **6.1.1** で特定したリスク及び機会に対して環境マネジメントシステムの中で行わなければならない取組みを，高いレベルで計画する．

The actions planned may include establishing environmental objectives (see 6.2) or may be incorporated into other environmental management system processes, either individually or in combination. Some actions may be addressed through other management systems, such as those related to occupational health and safety or business continuity, or through other business processes related to risk, financial or human resource management.

When considering its technological options, an organization should consider the use of best-available techniques, where economically viable, cost-effective and judged appropriate. This is not intended to imply that organizations are obliged to use environmental cost-accounting methodologies.

## A.6.2  Environmental objectives and planning to achieve them

Top management may establish environmental objectives at the strategic level, the tactical level or the operational level. The strategic level includes the highest levels of the organization and the environmental objectives can be applicable to the whole

計画した取組みには，環境目標の確立（**6.2** 参照）を含めても，又は，この取組みを他の環境マネジメントシステムプロセスに，個別に若しくは組み合わせて組み込んでもよい．これらの取組みは，労働安全衛生，事業継続などの他のマネジメントシステムを通じて，又はリスク，財務若しくは人的資源のマネジメントに関連した他の事業プロセスを通じて行ってもよい．

技術上の選択肢を検討するとき，組織は，経済的に実行可能であり，費用対効果があり，かつ，適切と判断される場合には，利用可能な最良の技術の使用を考慮することが望ましい．これは，組織に環境原価会計手法の使用を義務付けようとするものではない．

## A.6.2　環境目標及びそれを達成するための計画策定

トップマネジメントは，戦略的，戦術的又は運用的レベルで，環境目標を確立してもよい．戦略的レベルは，組織の最高位を含み，その環境目標は，組織全体に適用できる．戦術的及び運用的レベルは，組織内の特定の単位又は機能のための環境目標を含

organization. The tactical and operational levels can include environmental objectives for specific units or functions within the organization and should be compatible with its strategic direction.

Environmental objectives should be communicated to persons working under the organization's control who have the ability to influence the achievement of environmental objectives.

The requirement to "take into account significant environmental aspects" does not mean that an environmental objective has to be established for each significant environmental aspect, however, these have a high priority when establishing environmental objectives.

"Consistent with the environmental policy" means that the environmental objectives are broadly aligned and harmonized with the commitments made by top management in the environmental policy, including the commitment to continual improvement.

み得るもので，組織の戦略的な方向性と両立していることが望ましい．

　環境目標は，その達成に影響を及ぼす能力をもつ，組織の管理下で働く人々に伝達することが望ましい．

　"著しい環境側面を考慮に入れる"という要求事項は，それぞれの著しい環境側面に対して環境目標を確立しなければならないということではないが，環境目標を確立するときに，著しい環境側面の優先順位が高いということを意味している．

　"環境方針と整合している"とは，環境目標が，継続的改善へのコミットメントを含め，環境方針の中でトップマネジメントが行うコミットメントと広く整合し，調和していることを意味する．

Indicators are selected to evaluate the achievement of measurable environmental objectives. "Measurable" means it is possible to use either quantitative or qualitative methods in relation to a specified scale to determine if the environmental objective has been achieved. By specifying "if practicable", it is acknowledged that there can be situations when it is not feasible to measure an environmental objective, however, it is important that the organization is able to determine whether or not an environmental objective has been achieved.

For additional information on environmental indicators, see ISO 14031.

## A.7  Support
### A.7.1  Resources
Resources are needed for the effective functioning and improvement of the environmental management system and to enhance environmental performance. Top management should ensure that those with environmental management system responsibilities are supported with the necessary resources. Internal resources may be supplemented

指標は，測定可能な環境目標の達成を評価するために選定される．“測定可能な”という言葉は，環境目標が達成されているか否かを決定するための規定された尺度に対して，定量的又は定性的な方法のいずれを用いることも可能であるということを意味する．“実行可能な場合”と規定しているとおり，環境目標を測定することが実施可能でない状況もあり得ることが認識されている．しかし，重要なことは，環境目標が達成されているか否かを決定できるのは組織であるということである．

環境指標に関する更なる情報は，ISO 14031 に示されている．

## A.7 支援

### A.7.1 資源

資源は，環境マネジメントシステムの有効な機能及び改善のため，並びに環境パフォーマンスを向上させるために必要である．トップマネジメントは，環境マネジメントシステムの責任をもつ人々が必要な資源によって支援されていることを確実にすることが望ましい．内部資源は，外部提供者によって補完してもよい．

206                    ISO 14001

by (an) external provider(s).

Resources can include human resources, natural resources, infrastructure, technology and financial resources. Examples of human resources include specialized skills and knowledge. Examples of infrastructure resources include the organization's buildings, equipment, underground tanks and drainage system.

## A.7.2  Competence

The competency requirements of this International Standard apply to persons working under the organization's control who affect its environmental performance, including persons:

a)  whose work has the potential to cause a significant environmental impact;

b)  who are assigned responsibilities for the environmental management system, including those who:

   1)  determine and evaluate environmental impacts or compliance obligations;

   2)  contribute to the achievement of an environmental objective;

資源には，人的資源，天然資源，インフラストラクチャ，技術及び資金が含まれ得る．人的資源の例には，専門的な技能及び知識が含まれる．インフラストラクチャの資源の例には，組織の建物，設備，地下タンク及び排水システムが含まれる．

## A.7.2　力量

この規格における力量の要求事項は，次に示す人々を含む，環境パフォーマンスに影響を与える，組織の管理下で働く人々に適用される．

a)　著しい環境影響の原因となる可能性をもつ業務を行う人

b)　次を行う人を含む，環境マネジメントシステムに関する責任を割り当てられた人

1)　環境影響又は順守義務を決定し，評価する．

2)　環境目標の達成に寄与する．

3) respond to emergency situations;

4) perform internal audits;

5) perform evaluations of compliance.

### A.7.3  Awareness

Awareness of the environmental policy should not be taken to mean that the commitments need to be memorized or that persons doing work under the organization's control have a copy of the documented environmental policy. Rather, these persons should be aware of its existence, its purpose and their role in achieving the commitments, including how their work can affect the organization's ability to fulfil its compliance obligations.

### A.7.4  Communication

Communication allows the organization to provide and obtain information relevant to its environmental management system, including information related to its significant environmental aspects, environmental performance, compliance obligations and recommendations for continual improvement. Communication is a two-way process, in and out of the organization.

附属書 A（参考）

3)  緊急事態に対応する．
4)  内部監査を実施する．
5)  順守評価を実施する．

## A.7.3  認識

環境方針の認識を，コミットメントを暗記する必要がある又は組織の管理下で働く人々が文書化した環境方針のコピーをもつ，という意味に捉えないほうがよい．そうではなく，環境方針の存在及びその目的を認識することが望ましく，また，自分の業務が，順守義務を満たす組織の能力にどのように影響を与え得るかということを含め，コミットメントの達成における自らの役割を認識することが望ましい．

## A.7.4  コミュニケーション

コミュニケーションによって，組織は，著しい環境側面，環境パフォーマンス，順守義務及び継続的改善のための提案に関する情報を含む，環境マネジメントシステムに関連した情報を提供し，入手することが可能となる．コミュニケーションは，組織の中と外との双方向のプロセスである．

When establishing its communication process(es), the internal organizational structure should be considered to ensure communication with the most appropriate levels and functions. A single approach can be adequate to meet the needs of many different interested parties, or multiple approaches might be necessary to address specific needs of individual interested parties.

The information received by the organization can contain requests from interested parties for specific information related to the management of its environmental aspects, or can contain general impressions or views on the way the organization carries out that management. These impressions or views can be positive or negative. In the latter case (e.g. complaints), it is important that a prompt and clear answer is provided by the organization. A subsequent analysis of these complaints can provide valuable information for detecting improvement opportunities for the environmental management system.

Communication should:

附属書A（参考）

コミュニケーションプロセスを確立するとき，最も適切な階層及び機能とコミュニケーションを行うことを確実にするために，内部の組織構造を考慮することが望ましい．多くの異なる利害関係者のニーズを満たすために，単一のアプローチをとることが適切な場合もあれば，個々の利害関係者の特定のニーズに取り組むために，複数のアプローチをとることが必要になることもあり得る．

組織が受け付ける情報には，環境側面のマネジメントに関する特定の情報に対する利害関係者からの要望を含むこともあれば，組織がそのマネジメントを遂行する方法についての一般的な印象又は見解を含むこともある．これらの印象又は見解は，肯定的なものもあれば，否定的なものもあり得る．後者（例えば，苦情）の場合には，組織が迅速かつ明確な回答を行うことが重要である．その後に行うこれらの苦情の分析は，環境マネジメントシステムの改善の機会を発見するための貴重な情報を提供し得る．

コミュニケーションは，次の事項を満たすことが

a) be transparent, i.e. the organization is open in the way it derives what it has reported on;

b) be appropriate, so that information meets the needs of relevant interested parties, enabling them to participate;

c) be truthful and not misleading to those who rely on the information reported;

d) be factual, accurate and able to be trusted;

e) not exclude relevant information;

f) be understandable to interested parties.

For information on communication as part of managing change, see Clause A.1. For additional information on communication, see ISO 14063.

### A.7.5 Documented information

An organization should create and maintain documented information in a manner sufficient to ensure a suitable, adequate and effective environmental management system. The primary focus should be on the implementation of the environ-

望ましい.

a) 透明である．すなわち，組織が，報告した内容の入手経路を公開している．

b) 適切である．すなわち，情報が，関連する利害関係者の参加を可能にしながら，これらの利害関係者のニーズを満たしている．

c) 偽りなく，報告した情報に頼る人々に誤解を与えないものである．

d) 事実に基づき，正確であり，信頼できるものである．

e) 関連する情報を除外していない．

f) 利害関係者にとって理解可能である．


変更のマネジメントの一部としてのコミュニケーションに関する情報を，**A.1** に示す．コミュニケーションに関する更なる情報は，**JIS Q 14063** に示されている．


## A.7.5 文書化した情報

組織は，適切で，妥当で，かつ，有効な環境マネジメントシステムを確実にするために十分な方法で，文書化した情報を作成し，維持することが望ましい．文書化した情報の複雑な管理システムではなく，環境マネジメントシステムの実施及び環境パフ

mental management system and on environmental performance, not on a complex documented information control system.

In addition to the documented information required in specific clauses of this International Standard, an organization may choose to create additional documented information for purposes of transparency, accountability, continuity, consistency, training, or ease in auditing.

Documented information originally created for purposes other than the environmental management system may be used. The documented information associated with the environmental management system may be integrated with other information management systems implemented by the organization. It does not have to be in the form of a manual.

## A.8   Operation
### A.8.1   Operational planning and control
The type and extent of operational control(s) depend on the nature of the operations, the risks and

ォーマンスに，最も焦点を当てることが望ましい．

この規格の特定の箇条で要求されている文書化した情報のほかに，組織は，透明性，説明責任，継続性，一貫性，教育訓練，又は監査の容易性のために，追加の文書化した情報を作成することを選んでもよい．

もともとは環境マネジメントシステム以外の目的のために作成された文書化した情報を，用いてもよい．環境マネジメントシステムに関する文書化した情報は，組織が実施している他の情報マネジメントシステムに統合してもよい．文書化した情報は，マニュアルの形式である必要はない．

## A.8　運用
### A.8.1　運用の計画及び管理
運用管理の方式及び程度は，運用の性質，リスク及び機会，著しい環境側面，並びに順守義務によっ

opportunities, significant environmental aspects and compliance obligations. An organization has the flexibility to select the type of operational control methods, individually or in combination, that are necessary to make sure the process(es) is (are) effective and achieve(s) the desired results. Such methods can include:

a)   designing (a) process(es) in such a way as to prevent error and ensure consistent results;

b)   using technology to control (a) process(es) and prevent adverse results (i.e. engineering controls);

c)   using competent personnel to ensure the desired results;

d)   performing (a) process(es) in a specified way;

e)   monitoring or measuring (a) process(es) to check the results;

f)   determining the use and amount of documented information necessary.


The organization decides the extent of control needed within its own business processes (e.g. procurement process) to control or influence (an) outsourced process(es) or (a) provider(s) of products

て異なる．組織は，プロセスが，有効で，かつ，望ましい結果を達成することを確かにするために必要な運用管理の方法を，個別に又は組み合わせて選定する柔軟性をもつ．こうした方法には，次の事項を含み得る．

a) 誤りを防止し，矛盾のない一貫した結果を確実にするような方法で，プロセスを設計する．

b) プロセスを管理し，有害な結果を防止するための技術（すなわち，工学的な管理）を用いる．

c) 望ましい結果を確実にするために，力量を備えた要員を用いる．

d) 規定された方法でプロセスを実施する．

e) 結果を点検するために，プロセスを監視又は測定する．

f) 必要な文書化した情報の使用及び量を決定する．

　組織は，外部委託したプロセス若しくは製品及びサービスの提供者を管理するため，又はそれらのプロセス若しくは提供者に影響を及ぼすために，自らの事業プロセス（例えば，調達プロセス）の中で必

and services. Its decision should be based upon factors such as:

— knowledge, competence and resources, including:

　— the competence of the external provider to meet the organization's environmental management system requirements;

　— the technical competence of the organization to define appropriate controls or assess the adequacy of controls;

— the importance and potential effect the product and service will have on the organization's ability to achieve the intended outcome of its environmental management system;

— the extent to which control of the process is shared;

— the capability of achieving the necessary control through the application of its general procurement process;

— improvement opportunities available.

When a process is outsourced, or when products and services are supplied by (an) external provider(s), the organization's ability to exert control or influ-

要な管理の程度を決定することとなる．この決定は，次のような要因に基づくことが望ましい．

— 次を含む，知識，力量及び資源

  — 組織の環境マネジメントシステム要求事項を満たすための外部提供者の力量

  — 適切な管理を決めるため，又は管理の妥当性を評価するための，組織の技術的な力量

— 環境マネジメントシステムの意図した成果を達成する組織の能力の重要性，並びにその能力に対して製品及びサービスが与える潜在的な影響

— プロセスの管理が共有される程度

— 一般的な調達プロセスを適用することを通して必要な管理を達成する能力

— 利用可能な改善の機会

  プロセスを外部委託する場合，又は製品及びサービスが外部提供者によって供給される場合，管理する又は影響を及ぼす組織の能力は，直接的に管理す

ence can vary from direct control to limited or no influence. In some cases, an outsourced process performed onsite might be under the direct control of an organization; in other cases, an organization's ability to influence an outsourced process or external supplier might be limited.

When determining the type and extent of operational controls related to external providers, including contractors, the organization may consider one or more factors such as:

— environmental aspects and associated environmental impacts;

— risks and opportunities associated with the manufacturing of its products or the provision of its services;

— the organization's compliance obligations.

For information on operational control as part of managing change, see Clause A.1. For information on life cycle perspective, see A.6.1.2.

An outsourced process is one that fulfils all of the following:

るものから，限定された影響を与えるもの又は全く影響をもたないものまで，異なり得る．ある場合には，現場で実施される外部委託したプロセスは，組織の直接的な管理下にあることがある．別の場合には，外部委託したプロセス又は外部供給者に影響を及ぼす組織の能力は，限定されることもある．

請負者を含む外部提供者に関連する運用管理の方式及び程度を決定するとき，組織は，次のような一つ又は複数の要因を考慮してもよい．

— 環境側面及びそれに伴う環境影響

— その製品の製造又はそのサービスの提供に関連するリスク及び機会

— 組織の順守義務

変更のマネジメントの一部としての運用管理に関する情報を，**A.1** に示す．ライフサイクルの視点に関する情報を，**A.6.1.2** に示す．

外部委託したプロセスとは，次の全ての事項を満たすものである．

— it is within the scope of the environmental management system;
— it is integral to the organization's functioning;
— it is needed for the environmental management system to achieve its intended outcome;
— liability for conforming to requirements is retained by the organization;
— the organization and the external provider have a relationship where the process is perceived by interested parties as being carried out by the organization.

Environmental requirements are the organization's environmentally-related needs and expectations that it establishes for, and communicates to, its interested parties (e.g. an internal function, such as procurement; a customer; an external provider).

Some of the organization's significant environmental impacts can occur during the transportation, delivery, use, end-of-life treatment or final disposal of its product or service. By providing information, an organization can potentially prevent or miti-

## 附属書A（参考）

— 環境マネジメントシステムの適用範囲の中にある．

— 組織が機能するために不可欠である．

— 環境マネジメントシステムが意図した成果を達成するために必要である．

— 要求事項に適合することに対する責任を，組織が保持している．

— そのプロセスを組織が実施していると利害関係者が認識しているような，組織と外部提供者との関係がある．

　環境上の要求事項とは，組織が利害関係者［例えば，内部機能（調達など），顧客，外部提供者］に対して確立し，伝達する，環境に関連した組織のニーズ及び期待である．

　組織の著しい環境影響には，製品又はサービスの輸送，配送（提供），使用，使用後の処理又は最終処分の中で発生し得るものもある．情報を提供することによって，組織は，これらのライフサイクルの段階において，有害な環境影響を潜在的に防止又は

224                    ISO 14001

gate adverse environmental impacts during these
life cycle stages.

## A.8.2  Emergency preparedness and response

It is the responsibility of each organization to be
prepared and to respond to emergency situations
in a manner appropriate to its particular needs.
For information on determining emergency situations, see A.6.1.1.

When planning its emergency preparedness and
response process(es), the organization should consider:

a)  the most appropriate method(s) for responding
    to an emergency situation;

b)  internal and external communication process
    (es);

c)  the action(s) required to prevent or mitigate
    environmental impacts;

d)  mitigation and response action(s) to be taken
    for different types of emergency situations;

e)  the need for post-emergency evaluation to determine and implement corrective actions;

附属書 A（参考） 225

緩和することができる．


## A.8.2　緊急事態への準備及び対応

　組織独自のニーズに適切な方法で緊急事態に対して準備し，対応することは，それぞれの組織の責任である．緊急事態の決定に関する情報を，**A.6.1.1**に示す．


　緊急事態への準備及び対応のプロセスを計画するとき，組織は，次の事項を考慮することが望ましい．

**a)**　緊急事態に対処する最適な方法

**b)**　内部及び外部コミュニケーションプロセス

**c)**　環境影響を防止又は緩和するのに必要な処置

**d)**　様々な種類の緊急事態に対してとるべき緩和及び対応処置

**e)**　是正処置を決定し実施するための緊急事態後の評価の必要性

f) periodic testing of planned emergency response actions;

g) training of emergency response personnel;

h) a list of key personnel and aid agencies, including contact details (e.g. fire department, spillage clean-up services);

i) evacuation routes and assembly points;

j) the possibility of mutual assistance from neighbouring organizations.

## A.9 Performance evaluation

## A.9.1 Monitoring, measurement, analysis and evaluation

### A.9.1.1 General

When determining what should be monitored and measured, in addition to progress on environmental objectives, the organization should take into account its significant environmental aspects, compliance obligations and operational controls.

The methods used by the organization to monitor and measure, analyse and evaluate should be defined in the environmental management system, in order to ensure that:

**f)** 計画した緊急事態対応処置の定期的なテストの実施

**g)** 緊急事態に対応する要員の教育訓練

**h)** 連絡の詳細（例えば，消防署，流出物の清掃サービス）を含めた，主要な要員及び支援機関のリスト

**i)** 避難ルート及び集合場所

**j)** 近隣組織からの相互支援の可能性

## A.9 パフォーマンス評価

## A.9.1 監視，測定，分析及び評価

### A.9.1.1 一般

環境目標の進捗のほかに，監視し測定することが望ましいものを決定するとき，組織は，著しい環境側面，順守義務及び運用管理を考慮に入れることが望ましい．

監視，測定，分析及び評価のために組織が用いる方法は，次の事項を確実にするために，環境マネジメントシステムの中で定めることが望ましい．

228  ISO 14001

a) the timing of monitoring and measurement is coordinated with the need for analysis and evaluation results;

b) the results of monitoring and measurement are reliable, reproducible and traceable;

c) the analysis and evaluation are reliable and reproducible, and enable the organization to report trends.

The environmental performance analysis and evaluation results should be reported to those with responsibility and authority to initiate appropriate action.

For additional information on environmental performance evaluation, see ISO 14031.

## A.9.1.2 Evaluation of compliance

The frequency and timing of compliance evaluations can vary depending on the importance of the requirement, variations in operating conditions, changes in compliance obligations and the organization's past performance. An organization can use a variety of methods to maintain its knowledge

附属書 A（参考） 229

**a)** 監視及び測定のタイミングが，分析及び評価の結果の必要性との関係で調整されている．

**b)** 監視及び測定の結果が信頼でき，再現性があり，かつ，追跡可能である．

**c)** 分析及び評価が信頼でき，再現性があり，かつ，組織が傾向を報告できるようにするものである．

環境パフォーマンスの分析及び評価の結果は，適切な処置を開始する責任及び権限をもつ人々に報告することが望ましい．

環境パフォーマンス評価に関する更なる情報は，**ISO 14031** に示されている．

## A.9.1.2　順守評価

順守評価の頻度及びタイミングは，要求事項の重要性，運用条件の変動，順守義務の変化，及び組織の過去のパフォーマンスによって異なることがある．組織は，順守状況に関する知識及び理解を維持するために種々の方法を用いることができるが，全ての順守義務を定期的に評価する必要がある．

and understanding of its compliance status, however, all compliance obligations need to be evaluated periodically.

If compliance evaluation results indicate a failure to fulfil a legal requirement, the organization needs to determine and implement the actions necessary to achieve compliance. This might require communication with a regulatory agency and agreement on a course of action to fulfil its legal requirements. Where such an agreement is in place, it becomes a compliance obligation.

A non-compliance is not necessarily elevated to a nonconformity if, for example, it is identified and corrected by the environmental management system processes. Compliance-related nonconformities need to be corrected, even if those nonconformities have not resulted in actual non-compliance with legal requirements.

## A.9.2   Internal audit

Auditors should be independent of the activity being audited, wherever practicable, and should in

順守評価の結果，法的要求事項を満たしていないことが示された場合，組織は，順守を達成するために必要な処置を決定し，実施する必要がある．この場合，規制当局とやり取りし，法的要求事項を満たすための一連の処置について合意することが求められ得る．このような合意がなされた場合，それは順守義務となる．

不順守は，例えばそれが環境マネジメントシステムプロセスによって特定され，修正された場合は，必ずしも不適合にはならない．順守に関連する不適合は，その不適合が法的要求事項の実際の不順守には至らない場合であっても，修正する必要がある．

## A.9.2　内部監査

監査員は，実行可能な限り，監査の対象となる活動から独立した立場にあり，全ての場合において偏

232 ISO 14001

all cases act in a manner that is free from bias and conflict of interest.

Nonconformities identified during internal audits are subject to appropriate corrective action.

When considering the results of previous audits, the organization should include:

a)  previously identified nonconformities and the effectiveness of the actions taken;

b)  results of internal and external audits.

For additional information on establishing an internal audit programme, performing environmental management system audits and evaluating the competence of audit personnel, see ISO 19011. For information on internal audit programme as part of managing change, see Clause A.1.

### A.9.3   Management review

The management review should be high-level; it does not need to be an exhaustive review of detailed information. The management review topics need not be addressed all at once. The review may

附属書 A（参考）　　　　233

り及び利害抵触がない形で行動することが望ましい．

　内部監査において特定された不適合は，適切な是正処置をとる必要がある．

　前回までの監査の結果を考慮するに当たって，組織は，次の事項を含めることが望ましい．
a)　これまでに特定された不適合，及びとった処置の有効性
b)　内部監査及び外部監査の結果

　内部監査プログラムの確立，環境マネジメントシステムの監査の実施，及び監査要員の力量の評価に関する更なる情報は，**JIS Q 19011** に示されている．変更のマネジメントの一部としての内部監査プログラムに関する情報を，**A.1** に示す．

## A.9.3　マネジメントレビュー

　マネジメントレビューは，高いレベルのものであることが望ましく，詳細な情報の徹底的なレビューである必要はない．マネジメントレビューの項目は，全てに同時に取り組む必要はない．レビュー

take place over a period of time and can be part of regularly scheduled management activities, such as board or operational meetings; it does not need to be a separate activity.

Relevant complaints received from interested parties are reviewed by top management to determine opportunities for improvement.

For information on management review as part of managing change, see Clause A.1.

"Suitability" refers to how the environmental management system fits the organization, its operations, culture and business systems. "Adequacy" refers to whether it meets the requirements of this International Standard and is implemented appropriately. "Effectiveness" refers to whether it is achieving the desired results.

## A.10   Improvement
## A.10.1   General

は，一定の期間にわたって行ってもよく，また，役員会，運営会議のような，定期的に開催される管理層の活動の一部に位置付けることもできる．したがって，レビューだけを個別の活動として分ける必要はない．

利害関係者から受け付けた関連する苦情は，改善の機会を決定するために，トップマネジメントがレビューする．

変更のマネジメントの一部としてのマネジメントレビューに関する情報を，**A.1** に示す．

"適切（性）"（suitability）とは，環境マネジメントシステムが，組織，並びに組織の運用，文化及び事業システムにどのように合っているかを意味している．"妥当（性）"（adequacy）とは，この規格の要求事項を満たし，十分なレベルで実施されているかどうかを意味している．"有効（性）"（effectiveness）とは，望ましい結果を達成しているかどうかを意味している．

## A.10 改善
### A.10.1 一般

The organization should consider the results from analysis and evaluation of environmental performance, evaluation of compliance, internal audits and management review when taking action to improve.

Examples of improvement include corrective action, continual improvement, breakthrough change, innovation and re-organization.

## A.10.2 Nonconformity and corrective action

One of the key purposes of an environmental management system is to act as a preventive tool. The concept of preventive action is now captured in 4.1 (i.e. understanding the organization and its context) and 6.1 (i.e. actions to address risks and opportunities).

## A.10.3 Continual improvement

The rate, extent and timescale of actions that support continual improvement are determined by the organization. Environmental performance can be enhanced by applying the environmental manage-

組織は，改善のための処置をとるときに，環境パフォーマンスの分析及び評価からの結果，並びに順守評価，内部監査及びマネジメントレビューからの結果を考慮することが望ましい．

改善の例には，是正処置，継続的改善，現状を打破する変更，革新及び組織再編が含まれる．

## A.10.2　不適合及び是正処置

環境マネジメントシステムの主要な目的の一つは，予防的なツールとして働くことである．予防処置の概念は，この規格では，**4.1**（組織及びその状況の理解）及び **6.1**（リスク及び機会への取組み）に包含されている．

## A.10.3　継続的改善

継続的改善を支える処置の度合い，範囲及び期間は，組織によって決定される．環境パフォーマンスは，環境マネジメントシステムを全体として適用することによって，又は環境マネジメントシステムの

ment system as a whole or improving one or more of its elements.

一つ若しくは複数の要素を改善することによって，
向上させることができる．

# Annex B
## (informative)

## Correspondence between ISO 14001:2015 and ISO 14001:2004

Table B.1 shows the correspondence between this edition of this International Standard (ISO 14001:2015) and the previous edition (ISO 14001:2004).

### Table B.1 — Correspondence between ISO 14001:2015 and ISO 14001:2004

| ISO 14001:2015 | | ISO 14001:2004 | |
|---|---|---|---|
| Clause title | Clause number | Clause number | Clause title |
| Introduction | | | Introduction |
| Scope | 1 | 1 | Scope |
| Normative references | 2 | 2 | Normative references |
| Terms and definitions | 3 | 3 | Terms and definitions |
| Context of the organization (title only) | 4 | | |
| | | 4 | Environmental management system requirements (title only) |
| Understanding the organization and its context | 4.1 | | |
| Understanding the needs and expectations of interested parties | 4.2 | | |

# 附属書 B

## （参考）

## JIS Q 14001:2015 と JIS Q 14001:2004 との対応

表 **B.1** は，この規格（**JIS Q 14001**:2015）と旧規格（**JIS Q 14001**:2004）との対応を示している．

### 表 B.1—JIS Q 14001:2015 と JIS Q 14001:2004 との対応

| JIS Q 14001:2015 | | JIS Q 14001:2004 | |
|---|---|---|---|
| 箇条のタイトル | 箇条番号 | 箇条番号 | 箇条のタイトル |
| 序文 | | | 序文 |
| 適用範囲 | 1 | 1. | 適用範囲 |
| 引用規格 | 2 | 2. | 引用規格 |
| 用語及び定義 | 3 | 3. | 用語及び定義 |
| 組織の状況(タイトルだけ) | 4 | | |
| | | 4. | 環境マネジメントシステム要求事項（タイトルだけ） |
| 組織及びその状況の理解 | 4.1 | | |
| 利害関係者のニーズ及び期待の理解 | 4.2 | | |

242  ISO 14001

**Table B.1** (continued)

| ISO 14001:2015 | | ISO 14001:2004 | |
|---|---|---|---|
| Clause title | Clause number | Clause number | Clause title |
| Determining the scope of the environmental management system | 4.3 | 4.1 | General requirements |
| Environmental management system | 4.4 | 4.1 | General requirements |
| Leadership (title only) | 5 | | |
| Leadership and commitment | 5.1 | | |
| Environmental policy | 5.2 | 4.2 | Environmental policy |
| Organizational roles, responsibilities and authorities | 5.3 | 4.4.1 | Resources, roles, responsibility and authority |
| Planning (title only) | 6 | 4.3 | Planning (title only) |
| Actions to address risks and opportunities (title only) | 6.1 | | |
| General | 6.1.1 | | |
| Environmental aspects | 6.1.2 | 4.3.1 | Environmental aspects |
| Compliance obligations | 6.1.3 | 4.3.2 | Legal and other requirements |
| Planning action | 6.1.4 | | |
| Environmental objectives and planning to achieve them (title only) | 6.2 | 4.3.3 | Objectives, targets and programme(s) |
| Environmental objectives | 6.2.1 | | |
| Planning actions to achieve environmental objectives | 6.2.2 | | |
| Support (title only) | 7 | 4.4 | Implementation and operation (title only) |

附属書 B（参考）　　　243

## 表 B.1（続き）

| JIS Q 14001:2015 | | JIS Q 14001:2004 | |
|---|---|---|---|
| 箇条のタイトル | 箇条番号 | 箇条番号 | 箇条のタイトル |
| 環境マネジメントシステムの適用範囲の決定 | 4.3 | 4.1 | 一般要求事項 |
| 環境マネジメントシステム | 4.4 | 4.1 | 一般要求事項 |
| リーダーシップ（タイトルだけ） | 5 | | |
| リーダーシップ及びコミットメント | 5.1 | | |
| 環境方針 | 5.2 | 4.2 | 環境方針 |
| 組織の役割，責任及び権限 | 5.3 | 4.4.1 | 資源，役割，責任及び権限 |
| 計画（タイトルだけ） | 6 | 4.3 | 計画（タイトルだけ） |
| リスク及び機会への取組み（タイトルだけ） | 6.1 | | |
| 一般 | 6.1.1 | | |
| 環境側面 | 6.1.2 | 4.3.1 | 環境側面 |
| 順守義務 | 6.1.3 | 4.3.2 | 法的及びその他の要求事項 |
| 取組みの計画策定 | 6.1.4 | | |
| 環境目標及びそれを達成するための計画策定（タイトルだけ） | 6.2 | 4.3.3 | 目的，目標及び実施計画 |
| 環境目標 | 6.2.1 | | |
| 環境目標を達成するための取組みの計画策定 | 6.2.2 | | |
| 支援（タイトルだけ） | 7 | 4.4 | 実施及び運用（タイトルだけ） |

244 ISO 14001

**Table B.1** (continued)

| ISO 14001:2015 | | ISO 14001:2004 | |
|---|---|---|---|
| **Clause title** | Clause number | Clause number | **Clause title** |
| Resources | 7.1 | 4.4.1 | Resources, roles, responsibility and authority |
| Competence | 7.2 | 4.4.2 | Competence, training and awareness |
| Awareness | 7.3 | | |
| Communication (title only) | 7.4 | 4.4.3 | Communication |
| General | 7.4.1 | | |
| Internal communication | 7.4.2 | | |
| External communication | 7.4.3 | | |
| Documented information (title only) | 7.5 | 4.4.4 | Documentation |
| General | 7.5.1 | | |
| Creating and updating | 7.5.2 | 4.4.5 | Control of documents |
| | | 4.5.4 | Control of records |
| Control of documented information | 7.5.3 | 4.4.5 | Control of documents |
| | | 4.5.4 | Control of records |
| Operation (title only) | 8 | 4.4 | Implementation and operation (title only) |
| Operational planning and control | 8.1 | 4.4.6 | Operational control |
| Emergency preparedness and response | 8.2 | 4.4.7 | Emergency preparedness and response |
| Performance evaluation (title only) | 9 | 4.5 | Checking (title only) |
| Monitoring, measurement, analysis and evaluation (title only) | 9.1 | 4.5.1 | Monitoring and measurement |
| General | 9.1.1 | | |
| Evaluation of compliance | 9.1.2 | 4.5.2 | Evaluation of compliance |

附属書 B（参考）　　　　245

## 表 B.1 （続き）

| JIS Q 14001:2015 | | JIS Q 14001:2004 | |
|---|---|---|---|
| 箇条のタイトル | 箇条番号 | 箇条番号 | 箇条のタイトル |
| 資源 | 7.1 | 4.4.1 | 資源，役割，責任及び権限 |
| 力量 | 7.2 | 4.4.2 | 力量，教育訓練及び自覚 |
| 認識 | 7.3 | | |
| コミュニケーション（タイトルだけ） | 7.4 | 4.4.3 | コミュニケーション |
| 一般 | 7.4.1 | | |
| 内部コミュニケーション | 7.4.2 | | |
| 外部コミュニケーション | 7.4.3 | | |
| 文書化した情報（タイトルだけ） | 7.5 | 4.4.4 | 文書類 |
| 一般 | 7.5.1 | | |
| 作成及び更新 | 7.5.2 | 4.4.5 | 文書管理 |
| | | 4.5.4 | 記録の管理 |
| 文書化した情報の管理 | 7.5.3 | 4.4.5 | 文書管理 |
| | | 4.5.4 | 記録の管理 |
| 運用（タイトルだけ） | 8 | 4.4 | 実施及び運用（タイトルだけ） |
| 運用の計画及び管理 | 8.1 | 4.4.6 | 運用管理 |
| 緊急事態への準備及び対応 | 8.2 | 4.4.7 | 緊急事態への準備及び対応 |
| パフォーマンス評価（タイトルだけ） | 9 | 4.5 | 点検（タイトルだけ） |
| 監視，測定，分析及び評価（タイトルだけ） | 9.1 | 4.5.1 | 監視及び測定 |
| 一般 | 9.1.1 | | |
| 順守評価 | 9.1.2 | 4.5.2 | 順守評価 |

246 ISO 14001

**Table B.1** *(continued)*

| ISO 14001:2015 | | ISO 14001:2004 | |
|---|---|---|---|
| **Clause title** | **Clause number** | **Clause number** | **Clause title** |
| Internal audit (title only) | 9.2 | 4.5.5 | Internal audit |
| General | 9.2.1 | | |
| Internal audit pro-gramme | 9.2.2 | | |
| Management review | 9.3 | 4.6 | Management review |
| Improvement (title only) | 10 | | |
| General | 10.1 | | |
| Nonconformity and cor-rective action | 10.2 | 4.5.3 | Nonconformity, correc-tive action and preven-tive action |
| Continual improvement | 10.3 | | |
| Guidance on the use of this International Stan-dard | Annex A | Annex A | Guidance on the use of this International Stan-dard |
| Correspondence be-tween ISO 14001:2015 and ISO 14001:2004 | Annex B | | |
| | | Annex B | Correspondence be-tween ISO 14001:2004 and ISO 9001:2008 |
| Bibliography | | | Bibliography |
| Alphabetical index of terms | | | |

附属書 B（参考）　　　　　247

**表 B.1**（続き）

| JIS Q 14001:2015 | | JIS Q 14001:2004 | |
|---|---|---|---|
| 箇条のタイトル | 箇条番号 | 箇条番号 | 箇条のタイトル |
| 内部監査(タイトルだけ) | 9.2 | 4.5.5 | 内部監査 |
| 一般 | 9.2.1 | | |
| 内部監査プログラム | 9.2.2 | | |
| マネジメントレビュー | 9.3 | 4.6 | マネジメントレビュー |
| 改善（タイトルだけ） | 10 | | |
| 一般 | 10.1 | | |
| 不適合及び是正処置 | 10.2 | 4.5.3 | 不適合並びに是正処置及び予防処置 |
| 継続的改善 | 10.3 | | |
| この規格の利用の手引 | 附属書 A | 附属書 A | この規格の利用の手引 |
| JIS Q 14001:2015 と JIS Q 14001:2004 との対応 | 附属書 B | | |
| | | 附属書 B | JIS Q 14001:2004 と JIS Q 9001:2000 との対応 |
| 参考文献 | | | 参考文献 |
| 用語索引（五十音順） | | | |
| 用語索引（アルファベット順） | | | |

# Bibliography

[1] ISO 14004, *Environmental management systems — General guidelines on principles, systems and support techniques*

[2] ISO 14006, *Environmental management systems — Guidelines for incorporating ecodesign*

[3] ISO 14031, *Environmental management — Environmental performance evaluation — Guidelines*

[4] ISO 14044, *Environmental management — Life cycle assessment — Requirements and guidelines*

[5] ISO 14063, *Environmental management — Environmental communication — Guidelines and examples*

[6] ISO 19011, *Guidelines for auditing management systems*

[7] ISO 31000, *Risk management — Principles and guidelines*

# 参 考 文 献

〔ISO の Bibliography と JIS の参考文献は，それぞれの原文
において異なっているため，対訳となっていないことにご注意
ください.〕

[1] **JIS Q 0073** リスクマネジメント―用語
　　**注記** 対応国際規格：**ISO Guide 73**, Risk management―Vocabulary（IDT）

[2] **JIS Q 14004** 環境マネジメントシステム―
原則，システム及び支援技法の一般指針
　　**注記** 対応国際規格：**ISO 14004**, Environmental management systems―General guidelines on principles, systems and support techniques（IDT）

[3] **JIS Q 14006** 環境マネジメントシステム―
エコデザインの導入のための指針
　　**注記** 対応国際規格：**ISO 14006**, Environmental management systems―Guidelines for incorporating ecodesign（IDT）

[4] **JIS Q 14044** 環境マネジメント―ライフサ
イクルアセスメント―要求事項及び指針
　　**注記** 対応国際規格：**ISO 14044**, Envi-

[8]  ISO 50001, *Energy management systems — Requirements with guidance for use*

[9]  ISO Guide 73, *Risk management — Vocabulary*

ronmental management—Life cycle assessment—Requirements and guidelines（IDT）

[5] **JIS Q 14063** 環境マネジメント—環境コミュニケーション—指針及びその事例

**注記** 対応国際規格：**ISO 14063**, Environmental management—Environmental communication—Guidelines and examples（IDT）

[6] **JIS Q 19011** マネジメントシステム監査のための指針

**注記** 対応国際規格：**ISO 19011**, Guidelines for auditing management systems（IDT）

[7] **JIS Q 31000** リスクマネジメント—原則及び指針

**注記** 対応国際規格：**ISO 31000**, Risk management—Principles and guidelines（IDT）

[8] **JIS Q 50001** エネルギーマネジメントシステム—要求事項及び利用の手引

**注記** 対応国際規格：**ISO 50001**, Energy management systems—Requirements with guidance for use（IDT）

ISO 14001

[9] **ISO 14031**, Environmental management
—Environmental performance evaluation—
Guidelines

# 索引 （五十音順）

## 【い】

意図した成果　intended outcome　　157

## 【え】

影響　effect　　153

## 【お】

汚染の予防　prevention of pollution　　61

## 【か】

該当する場合には，必ず　applicable　　151
外部委託する　outsource　　71
外部提供者　external provider　　155
確実にする　ensure　　153
確保する　ensure　　153
環境　environment　　55
環境影響　environmental impact　　59, 153
環境状態　environmental condition　　59
環境側面　environmental aspect　　57
環境パフォーマンス　environmental performance
　　81
環境方針　environmental policy　　51

環境マネジメントシステム　environmental
　management system　51
環境目標　environmental objective　61, 159
監査　audit　73
監視　monitoring　79

## 【け】

継続的　continual　153
継続的改善　continual improvement　77
決定する　determine　157

## 【こ】

考慮する　consider　151
考慮に入れる　take into account　151

## 【し】

〜し得る　can　43
〜してもよい　may　43
〜しなければならない　shall　43
指標　indicator　79
順守義務　compliance obligation　65, 155

## 【す】

ステークホルダー　stakeholder　153
〜することができる　can　43
〜することが望ましい　should　43

## 【せ】

是正処置　corrective action　　77

## 【そ】

測定　measurement　　81
組織　organization　　53
組織の管理下で働く人（又は人々）
　　person(s) doing work under its control　　157

## 【て】

適合　conformity　　75
適切な　appropriate　　151
適用される，適用できる　applicable　　151
〜できる　can　　43

## 【と】

特定する　identify　　157
トップマネジメント　top management　　53

## 【は】

パフォーマンス　performance　　81

## 【ひ】

必要に応じて　appropriate　　151

## 【ふ】

不適合　nonconformity　　75

プロセス　process　　73

文書化した情報　documented information　　69, 155

### 【ま】

マネジメントシステム　management system　　49

### 【も】

目的　objective　　59
目標　objective　　59
目標　target　　159

### 【ゆ】

有効性　effectiveness　　79

### 【よ】

要求事項　requirement　　63

### 【ら】

ライフサイクル　life cycle　　71

### 【り】

利害関係者　interested party　　55, 153
力量　competence　　69
リスク　risk　　65
リスク及び機会　risks and opportunities　　67

# 索引 （アルファベット順）

## 【A】

applicable 該当する場合には，必ず／
　　　　　　適用される，適用できる　150
appropriate 適切な，必要に応じて　150
audit 監査　72

## 【C】

can 〜し得る，〜することができる，〜できる
　42
competence 力量　68
compliance obligations 順守義務　64, 154
conformity 適合　74
consider 考慮する　150
continual 継続的　152
continual improvement 継続的改善　76
corrective action 是正処置　76

## 【D】

determine 決定する　156
documented information 文書化した情報　68,
　154

## 【E】

effect　影響　　152
effectiveness　有効性　　78
ensure　確実にする，確保する　　152
environment　環境　　54
environmental aspect　環境側面　　56
environmental condition　環境状態　　58
environmental impact　環境影響　　58, 152
environmental management system　環境マネジメ
　ントシステム　　50
environmental objective　環境目標　　60, 158
environmental performance　環境パフォーマンス
　80
environmental policy　環境方針　　50
external provider　外部提供者　　154

## 【I】

identify　特定する　　156
indicator　指標　　78
intended outcome　意図した成果　　156
interested party　利害関係者　　54, 152

## 【L】

life cycle　ライフサイクル　　70

## 【M】

management system　マネジメントシステム　　48

may　〜してもよい　42

measurement　測定　80

monitoring　監視　78

## 【N】

nonconformity　不適合　74

## 【O】

objective　目的，目標　58

organization　組織　52

outsource　外部委託する　70

## 【P】

performance　パフォーマンス　80

person(s) doing work under its control
　組織の管理下で働く人（又は人々）　156

prevention of pollution　汚染の予防　60

process　プロセス　72

## 【R】

requirement　要求事項　62

risk　リスク　64

risks and opportunities　リスク及び機会　66

## 【S】

shall　〜しなければならない　42

should　〜することが望ましい　42

stakeholder　ステークホルダー　152

## 【T】

take into account　考慮に入れる　　150
target　目標　　158
top management　トップマネジメント　　52

**対訳 ISO 14001:2015（JIS Q 14001:2015）**
**環境マネジメントの国際規格　［ポケット版］**

2016 年 2 月 26 日　　第 1 版第 1 刷発行
2025 年 4 月 18 日　　　　　第 8 刷発行

編　　者　一般財団法人 日本規格協会

発 行 者　朝日　弘

発 行 所　一般財団法人 日本規格協会

　　　　　〒 108-0073　東京都港区三田 3 丁目 11-28 三田 Avanti
　　　　　https://www.jsa.or.jp/
　　　　　振替　00160-2-195146

製　　作　日本規格協会ソリューションズ株式会社
印 刷 所　株式会社ディグ

© Japanese Standards Association, et al., 2016　　Printed in Japan
ISBN978-4-542-40267-6

- 当会発行図書，海外規格のお求めは，下記をご利用ください．
  JSA Webdesk（オンライン注文）：https://webdesk.jsa.or.jp/
  電話：050-1742-6256　E-mail：csd@jsa.or.jp

## 図書のご案内

### ISO 14001:2015 (JIS Q 14001:2015)
### 要求事項の解説

ISO/TC 207/SC 1 日本代表委員　　ISO/TC 207/SC 1 日本代表委員
環境管理システム小委員会委員長　　環境管理システム小委員会委員
吉田敬史　　・　　奥野麻衣子　共著
A5 判・322 ページ　　定価：4,180 円（本体 3,800 円＋税 10％）

### リスク及び機会　実践ガイド

　　ISO 14001 を中心に

吉田敬史　著
A5 判・188 ページ　　定価 2,750 円（本体 2,500 円＋税 10％）

### 効果の上がる ISO 14001:2015
### 実践のポイント

吉田敬史　著
A5 判・206 ページ　　定価 2,970 円（本体 2,700 円＋税 10％）

### ［2015 年改訂対応］
### やさしい ISO 14001 (JIS Q 14001)
### 環境マネジメントシステム入門

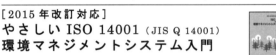

吉田敬史　著
A5 判・134 ページ　　定価 1,650 円（本体 1,500 円＋別 10％）

### 見るみる ISO 14001

**イラストとワークブックで要点を理解**

寺田和正・深田博史・寺田　博　著
A5 判・120 ページ　　定価 1,100 円（本体 1,000 円＋税 10％）

### ［2015 年版対応］
### 活き活き ISO 14001
　　―本音で取り組む環境活動―

国府保周　著
新書判・188 ページ　　定価 1,540 円（本体 1,400 円＋税 10％）

日本規格協会　　https://webdesk.jsa.or.jp/